T0029096

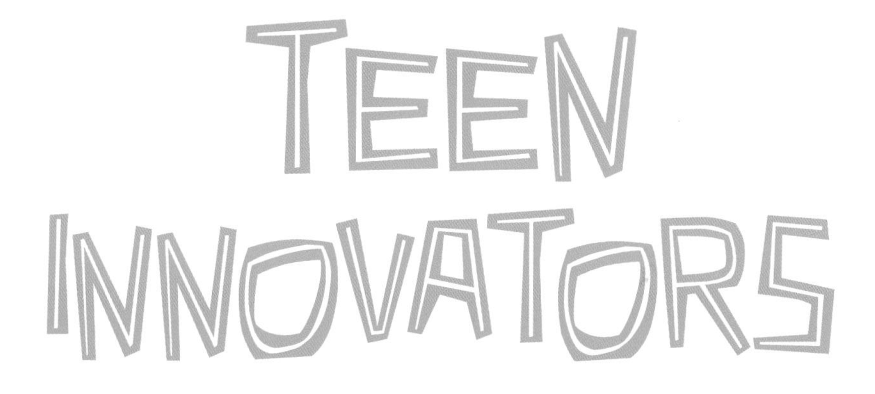

Teen Innovators

Nine YOUNG PEOPLE Engineering a BETTER WORLD with CREATIVE INVENTIONS

For Austin Veseliza, ingenious, hard-working, socially conscious, and compassionate

Text copyright © 2022 by Fred Estes

All rights reserved. No part of this book may be reproduced, stored in a retrieval system, or transmitted in any form or by any means—electronic, mechanical, photocopying, recording, or otherwise—without the prior written permission of Lerner Publishing Group, Inc., except for the inclusion of brief quotations in an acknowledged review.

Zest Books™
An imprint of Lerner Publishing Group, Inc.
241 First Avenue North
Minneapolis, MN 55401 USA

For reading levels and more information, look up this title at www.lernerbooks.com.
Visit us at zestbooks.net.

Cover and interior illustrations by Kavel Rafferty.

Designed by Athena Currier.
Main body text set in Tw Cen MT Std Light.
Typeface provided by Monotype Typography.

Library of Congress Cataloging-in-Publication Data

Names: Estes, Fred, 1950– author. | Rafferty, Kavel, illustrator.
Title: Teen innovators : nine young people engineering a better world with creative inventions / Fred Estes ; [illustrated by] Kavel Rafferty.
Description: Minneapolis : Zest Books, [2022] | Includes bibliographical references and index. | Audience: Ages: 10–15 | Audience: Grades: 7–9 | Summary: "Teen Innovators celebrates the determination and ingenuity of ten young people who created their own original inventions. From water testing to windmills, these youth use unique methods to overcome real world problems" —Provided by publisher.
Identifiers: LCCN 2021049294 (print) | LCCN 2021049295 (ebook) | ISBN 9781728417189 (library binding) | ISBN 9781728417219 (paperback) | ISBN 9781728445458 (ebook)
Subjects: LCSH: Inventors—Biography—Juvenile literature. | Youth—Biography—Juvenile literature.
Classification: LCC T39 .E78 2022 (print) | LCC T39 (ebook) | DDC 609.2—dc23/eng/20211118

LC record available at https://lccn.loc.gov/2021049294
LC ebook record available at https://lccn.loc.gov/2021049295

Manufactured in the United States of America
1-48972-49232-3/25/2022

TABLE OF CONTENTS

FOREWORD

Inventors have always fascinated me, because I grew up in a household where creativity was encouraged, and the materials we needed to create lay everywhere. I remember amateur radio kits, a photo lab, garden tools, woodworking tools, a sewing machine, and old how-to-do-it books to help us with our projects. I often worry about the kids who grow up without all the time and books and weird stuff at hand to invent things. It's no wonder I became a psychologist who studies innovators as well as provides guidance to young people who want to live creative lives. I've had the good fortune to be a professor who not only reads and writes about creative people but also gets to travel all over the world, meeting people who invent and the teachers who make classrooms into creative spaces like my own childhood home.

That's how I met Fred Estes, the writer of this book. For many years, I worked with the teachers of Nueva School, an amazing school that was then a beautiful, big old house in the woods overlooking San Francisco bay. I remember the "forts" the kids were able to build in the forest under the gentle supervision of teachers; the violin music ringing through the courtyard; and the quiet bustle of classrooms with children actively writing, painting, and exploring together. I'll never forget the first glimpse of Fred in his classroom, where he and his students were examining leaves for their veins, their coloring, their role in our world. There he was, walking briskly among groups of students, adjusting a microscope for some, exclaiming at the discovery of insect eggs on the underside of a leaf for another group, or pausing to quickly give a lesson that drew everybody's attention about what makes photosynthesis stop in the fall. How could a student who was exposed to such information and hands-on learning not be inspired to create?

When I learned he was writing a book about young innovators for young adults, I was thrilled, because I knew he would show us what innovation looked like for the kids themselves and the teachers who encouraged them. That's exactly what this book does. I meant to read it a chapter at a time, but instead ended up devouring it in one sitting. Each story reads like the best kind of mystery and adventure. How will this boy with no money use all the junk laying around his village to build a working windmill to power homes and draw water from the ground? How can this girl possibly find a way to fit all the devices she needs to purify water into one lightweight bottle, working with, of all things, concrete? How will this motley crew of undocumented kids at one of the poorest schools in Arizona figure out a way to make an underwater robot for national competitions while escaping the constant threat of deportation?

In each case, the kids *do* solve the mystery, while having an adventure of a lifetime. You hear all the time that anybody can be creative with enough hard work, but that is so misleading. Hard work sounds like having to do meaningless, boring stuff under the stern eye of some awful boss. This hard work, that indeed anybody can do, is actually a continuous set of engaging, independent adventures in learning how things work, and seeing ideas burst into being at each step of the way. This book will ignite your desire to seek answers to the big problems around you by gathering up some materials, a team of like-minded friends, and a lively mentor like Fred so you can make things that change the world.

Barbara Kerr, Distinguished Professor and Co-Director,
Center for Creativity and Entrepreneurship at the University of Kansas

INTRODUCTION

THE MUSIC FANS AND THE TURTLE RESEARCHER

Daniel and Martin traveled to Costa Rica with their tenth-grade class in the spring of 2015. Like their classmates, they were eager to learn more about leatherback turtles. Their biology class back in California had studied these endangered animals and planned this special field trip. The trip took them to a conservation preserve in Costa Rica where the students would live and study alongside the biologists, getting hands-on experience with research. Daniel and Martin planned to have fun with their classmates. They did not expect to discover a way to improve the research of a biologist at the station.

During the day, the students lived like marine biologists. Small teams followed different biologists around and assisted them in the ongoing research mission of the station. They counted turtles on the beaches for the census, visited known nests to track the progress of the hatching eggs, and recorded data about the hatchlings when they emerged. The biologists taught the students to identify different sea turtle species and about the habits and life cycles of these creatures that have swum the oceans since dinosaurs roamed Earth. Biologists estimate there is only one turtle for every thousand turtles that existed back then. All the species of great sea turtles are at risk of extinction. Daniel, Martin, and their classmates learned that marine biologists hope to preserve the great sea turtles by learning more about them. Maybe more knowledge and informed conservation could save the sea turtles from extinction.

Ancient Turtles

The great sea turtles are among the oldest reptiles on Earth, with origins in the late Jurassic period about 150 million years ago. Leatherbacks, the largest of these sea turtles, migrate across vast stretches of ocean, from Chile to Japan to Alaska and from the Caribbean to Nova Scotia to the eastern coast of Africa. Adult turtles may grow up to 9 feet (2.7 m) long and weigh up to 1,500 pounds (680 kg). They play a vital role in the ecosystems of the ocean and their nesting beaches. Leatherbacks eat vast quantities of jellyfish, over 400 pounds (181 kg) each day, and graze on seagrass. They bring nutrients to beaches where they nest, helping plants grow and preventing beach erosion. In doing all of this, leatherbacks play a key role in helping balance food webs in the ocean and on land.

Leatherbacks differ from other species of sea turtles because they do not have a hard shell. Instead, a tough, leathery skin, dark gray or black with white spots, covers their back, or carapace. The scientific name for leatherbacks (*Dermochelys coriacea*) means "leathery-skinned turtle." These flexible skins help leatherbacks dive as deep as 4,000 feet (1,219 m) in the

ocean, deeper than any other sea turtle. The water pressure at this depth, about 1,800 pounds per square inch (12.4 MPa), might crack a rigid shell. The leatherbacks have no rigid breastbone so their flexible shells compress as they dive. Humans with scuba gear can only dive about 250 feet (76 m), where the pressure is about 120 pounds per square inch (827 kPa), compared to the normal atmospheric pressure at sea level of about 15 pounds per square inch (103 kPa).

The migrations of a female leatherback can last from one to three years before she mates and returns to her nesting beach. She may lay over one hundred eggs in a nest she digs in the sand. Leatherbacks are pelagic—they spend most of their life in the open ocean. Leatherbacks do not live in groups but swim the ocean alone. This is another reason why finding mates is more difficult for them.

About sixty-five million years ago, an asteroid smashed into our planet near Mexico's Yucatán Peninsula with an impact as strong as 100 million tons (91 million t) of TNT. This collision triggered a dramatic change in Earth's climate and a mass extinction called the K-T event that wiped out the dinosaurs and about 80 percent of life on Earth. The ancestors of the leatherbacks somehow survived this cataclysm. But can they also survive the impact of people on their environment?

In the evening, the biologists described their research to the students. One biologist explained she was working on the puzzle of how to restore the gender balance in the turtle population, which heavily skews toward females. Unlike those of mammals and most other animals, sea turtles' genders are determined by the temperature of the nest as the young turtle develops in the egg. When the temperature is below 82°F (28°C), the hatchling will be male. If the temperature is above 88°F (31°C), then the turtle will be female. Both climate change and human development near nesting beaches raise the temperature of the beaches, so around 85 percent of the hatchlings are female. A more equal gender balance will help turtle populations grow, as it will be easier for turtles to find mates. Leatherbacks spend most of their lives swimming alone in the ocean, rarely encountering another leatherback, and adult males never return to land. Changing the gender ratio will make it more likely for a pair of potential mates to meet.

To make progress on this research, biologists need a simple way to determine the gender of the turtle hatchlings. All turtle hatchlings look alike on the outside. Researchers could not tell males from females by sight alone. This biologist told the students she thought the sounds the hatchlings make might be the key. She put recording devices near leatherback turtle hatching sites. Then she recorded the voices of the newborn leatherbacks. This scientist showed the students how she analyzed the turtle voices and explained her conclusions so far.

At the end of her talk, almost as a casual comment, she mentioned how her research required many boring and wasted hours. She explained that she listened to long audio recordings, searching for a few minutes of hatchling vocalizations for her to analyze. Her tapes recorded hours and hours of pleasant natural background sounds. She heard ocean waves on the beach and occasional beachgoers while she waited for eggs to hatch. She couldn't just fast-forward through the tapes because the hatchling vocalizations were too easy to miss.

The National Oceanic and Atmospheric Administration estimated in 2020 that leatherback sea turtles have a high extinction risk. Conservation efforts are crucial for preserving this species.

As they listened to her, Daniel and Martin thought they could help her. They both loved music and knew about Shazam, an app that could identify songs just by hearing a short sample. To develop the app, four young entrepreneurs from California had invented a way to make an "audio fingerprint" of any song and stored those fingerprints in a database. Shazam then simply matches samples to the audio fingerprints.

Daniel and Martin realized that they could identify leatherback turtle vocalizations using the same idea. They asked the scientist to identify the sound passages she was looking for. Then they used the sound technology to create audio fingerprints of the turtle song and put these fingerprints in a database. Daniel and Martin worked together to code software that would quickly scan hours of recordings. When the software found the turtle song on a recording, it placed an electronic tag to mark the spot.

Then the biologist could simply find these tags to hear the sounds of the hatchlings. Daniel and Martin's hatchling recognition software allows her to work faster and boosts her research output.

This book tells the stories of teens like Daniel and Martin who discover and invent, making important contributions in the fields of science and technology. I wrote this book because I could not find one like it, a book of stories like Martin and Daniel's. Before I knew about Daniel and Martin, however, I knew about Jack.

Finding Jack

My middle school science students sat stone silent—not the norm for my energetic and inquisitive class. As I told the story of fifteen-year-old Jack Andraka, they listened, spellbound.

After losing a close friend of his family to pancreatic cancer, Jack vowed to help others with this disease. His two-year quest resulted in a new diagnostic test that is faster and more accurate than the current one. It costs pennies instead of hundreds of dollars. Even better, Jack's test finds cancer much sooner. With early treatment, doctors can save up to 95 percent of pancreatic cancer patients.

> Everyone says that young people are the future. Well, we're here now. And we can help.
>
> —*Gitanjali Rao*

Did the idea for this test strike Jack in a flash of intuition? Much of Jack's discovery, after all, depended on science taught in middle school. Our class had just studied some of the principles of electricity Jack used. Study of the human immune system and antibodies, which he also used, came later in the curriculum. So, did he just make a lucky guess based on his schoolwork?

No, there was hard work. He spent an entire summer vacation combing a medical database to find a biomarker of pancreatic cancer. He read research papers and taught himself how to make sense of them. His parents and others supported him, while passion and patience kept him going. All this work prepared him for a clever and creative insight.

Then the real work began. He labored many late nights and long weekends in a lab while still going to high school full time. Jack read more research papers and learned the lab skills he needed. Developing the test was a slow, careful process. He continued until it worked.

The story of his success captivated us all—both my students and me. In his first TED Talk, Jack offers a challenge to viewers. "You could change the world . . . just imagine what you could do." I decided to find more stories like Jack's.

Needing a Book

But where were the books with these stories? Many books covered the lives and works of outstanding past scientists, engineers, mathematicians, and technologists. These were adults from long ago. Where were the stories of young people who had big ideas?

And how did these talented scientists think of their ideas? What led to their discoveries? In the biographies, their breakthroughs seemed like dramatic thunderbolts, unexpected and mysterious. How are brilliant discoveries made? In his TED Talk, Jack spoke about his sources of inspiration and his careful preparation. Were there patterns and practices to learn from?

Great ideas in science also build on earlier, simpler ideas. Good ideas combine with other good ideas to make better ideas. Ideas from different people, places, and backgrounds improve inventions. Great inventions never come from nowhere but are the creative synthesis of other ideas that came before them.

What This Book Is About

This book is about teen inventors and their inventions. Each chapter tells a story of discovery.

Jack Andraka thought of a new test for cancer while distracted in biology class. He was reading an article under his desk about how carbon atoms could form small tubes. His teacher was lecturing on how antibodies fight disease. Jack saw a solution to his research problem by linking these two ideas.

Gitanjali Rao, a middle school student, learned about Michigan children sickened by drinking water poisoned with lead. The city had tried to save money by using river water contaminated with industrial waste. While trying to treat the polluted water, the city ended up pumping water with toxic amounts of lead into people's homes. Gitanjali found a better way to test for lead in water by combining what she knew about electricity with what she learned about chemistry.

William Kamkwamba grew up in a farming family in the developing nation of Malawi. Periodic droughts brought famines to this small eastern African country. Forced to drop out of school because his family could not afford tuition, William taught himself the science he loved by reading in the library. Learning how windmills could generate electrical power, he saw a way to bring water to his family's parched fields. Despite the skepticism and ridicule of others, William, with the support of his family, persisted in his quest to build a windmill. His success changed life in his village forever.

Austin Veseliza developed a "talking glove" for people with speech difficulties. These people can use Austin's glove to communicate with other people who do not know sign language. After watching a woman try unsuccessfully to communicate in a store, Austin saw this need. When his gamer friends told him about a new joystick built into a glove, he envisioned a solution.

On a family trip to India, Deepika Kurup saw young children drinking contaminated water from puddles on the side of the road. Her parents always insisted the family drink only bottled water while traveling in India, so Deepika was puzzled. She learned that in many places in India and around the world, clean water is not always available. Many people are sickened or die from drinking contaminated water. Her invention of a method for individuals and families to purify their own water quickly and cheaply could potentially save millions of lives.

Cristian Arcega, Lorenzo Santillan, Oscar Vasquez, and Luis Aranda, four undocumented immigrant students from Phoenix, built an underwater robot. They met when two teachers formed a robotics club at their underfunded urban high school. These teachers gave them a challenge. Using spare parts, hard work, and ingenuity, they won a national robotics competition.

These teens saw problems and found answers. These trailblazers used science and their ingenuity to make the world better. These are the stories for my students and for anyone interested in young inventors and discoverers.

In addition to these main stories, sidebars in the chapters go into greater depth about the science and technology each discoverer uses and about various global and social issues. At the end of the book, you'll also find a glossary that defines any unfamiliar terminology, as well as a list of resources, including books, websites, and videos, for going further into many of the topics covered in each chapter.

Everyday Miracles

Remarkable events of creativity happen all the time, every single day. Some are big, like these stories of teen inventors. Some are small, like a clever hack you devise to fix something around your home. Maybe it's something simple, like realizing that using a fork to hold a nail while you hammer it will protect your fingers. Maybe it's something more elaborate, like rigging an Arduino microcontroller to water house plants automatically while you travel. Sharing these stories acknowledges that we are all inventive and creative. Opportunities to improve the world are everywhere. By remembering this, we increase the chances we will seize more of these opportunities.

We all have great ideas. While invention is not easy, it is possible and accessible for everyone, regardless of age, background, or circumstance. Maybe your story will be in the next edition of this book.

1

JACK ANDRAKA AND HIS IMPROVED CANCER TEST

Uncle Ted has cancer? Thirteen-year-old Jack Andraka couldn't believe what his mother told him as they walked near home one day after school.

Uncle Ted, a close friend of the family, supported and mentored Jack for years. Uncle Ted took him crabbing in the Chesapeake Bay near their home in Maryland. Uncle Ted listened to Jack's troubles and showed him shortcuts in math problems. And Uncle Ted always took an interest in Jack's science projects. When everyone else in the family seemed to focus on Luke, Jack's tall, athletic, brilliant, and charismatic older brother, Uncle Ted paid attention to Jack. *How could big, outdoorsy Uncle Ted be sick?*

Was pancreatic cancer serious? Jack wondered. He asked his mother if Uncle Ted would be OK. Looking as if she were trying to stay calm and project confidence, she told him that Uncle Ted was getting great medical care and that his doctors were doing everything they could to make him well.

As soon as they got home, Jack recalls, "I went to my room, closed the door, buried my head under the covers, and cried."

One More Thing

This was just one more bad thing in Jack's life. His two best friends had moved away. At school, Jack was the target of bullies. They taunted him in the halls, and everyone seemed to shun him in the cafeteria. Jack recalls, "Remember that kid in middle school who had big, thick glasses, wore braces, and was always raising his hand in class? Yeah, that was me." Jack's success with his winning science fair projects only seemed to make it worse.

Around the start of seventh grade, Jack also began to realize that he was gay. He knew he was attracted to boys and not girls. *What is going on with me?* Jack thought. "I took my feelings, and I locked them deep in a vault, and I did my very best to forget about them entirely. I stayed focused on science." But the bullying, the name-calling, and the bumping and shoving only got worse.

Through all the torture of middle school, Uncle Ted listened to Jack's troubles and gave unwavering support. "Middle school can be a rough time," he told Jack, "but things will get better in high school. You're going to do great things one day. I just know it." And now Uncle Ted, one of his main champions, was seriously ill. Fatally ill?

Pancreatic cancer moves quickly once it begins to spread in the body. Uncle Ted entered the hospital and weakened quickly. When Jack first visited, Uncle Ted looked much the same as always, a large, sturdy man with thinning brown hair. A few weeks later, besieged at school after coming out as gay, Jack felt a strong need to see Uncle Ted. Uncle Ted always knew just the right thing to say, and Jack valued his wise advice.

Uncle Ted's thin and frail appearance, however, shocked Jack. He seemed to have aged a couple of decades in a few weeks. When they hugged, Jack felt Uncle Ted's bones through the thin hospital gown. Seeing Uncle Ted's condition, Jack dropped the idea of sharing his problems. He didn't want to worry Uncle Ted.

They talked instead about Jack's next science project using bacteria to detect water pollution, which Jack knew he would enjoy hearing about.

Uncle Ted's Death and the Collapsing World

One spring afternoon, not long after this hospital visit, Jack came home from school to find his mother waiting for him. With tears in her eyes, she told Jack that Uncle Ted had died.

Jack had thought that a miracle would occur somehow. Uncle Ted would recover and be his old self again. There were so many things he had wanted to ask his Uncle Ted, to tell his Uncle Ted. But it was too late.

Jack's world collapsed. "I felt my stomach crash down to my feet. In those moments that followed, I felt as though I was looking in on my life from afar."

With all that was going wrong, Uncle Ted had been an anchor in his chaotic life. Jack became deeply depressed. Riding home from the memorial service, Jack stared out the window with unseeing eyes, wondering when he would wake from this never-ending nightmare.

Feeling alone, rejected, and depressed, Jack attempted suicide one day in the school bathroom. "If you've never felt depression," Jack says, "it's hard to explain. It was as if a massive blanket of hopelessness had draped over me. It was heavy, and no matter how hard I tried, I couldn't shake it off." After repeatedly stabbing his wrists with a pencil, Jack realized he wasn't succeeding even at this. With bloody arms, he walked into the hallway. Another student saw him and brought counselors on the run. "The next thing I remember, my parents were there at school. After that, it all goes dark."

After his suicide attempt, Jack got professional counseling and slowly began to make his way out of his depression. As he did, his thoughts turned to Uncle Ted. What would Uncle Ted say about all of this? Uncle Ted had struggled for life, fighting a battle with his pancreatic cancer. Uncle Ted had always told him how much Jack had to look forward to, how much potential Jack had. How could he throw that away?

Jack realized that if Uncle Ted were with him, Uncle Ted would ask him about his next science project and begin moving him forward along that path. Jack realized, "More than anything, I really wanted to get back to what I loved—science." He needed to start working on his next project and leave the rest behind.

Committing to Do Something

As Jack progressed in his recovery, he knew he needed to confront the loss of Uncle Ted. More than anything, he wanted to understand why Uncle Ted had died. How had Uncle Ted gotten pancreatic cancer? Why couldn't his doctors save him?

He learned that Uncle Ted's doctors had detected his cancer at too late a stage. The cancer had spread throughout his body, or metastasized. It was too late for surgery, chemotherapy, and radiation. Jack found out that doctors rarely found pancreatic cancer in time to save the patient.

"And that's when I had an idea," Jack says. "Maybe, just maybe, I could find a cure for pancreatic cancer." Looking back, Jack realizes what an ambitious project that was. Or overambitious, considering all the brilliant, highly trained medical researchers working on this problem with no success yet. "Whether it was youthful exuberance or even unbridled stupidity, I didn't know for sure, but for whatever reason, I was all in. Turns out I was the only one."

Both of Jack's parents were skeptical of Jack's plan, and his father asked if he was aiming too high. Jack knew he needed his parents behind him if he were to succeed. His parents, meanwhile, worried about Jack investing too heavily in a seemingly impossible project. If he failed, would it push him over the edge again?

Jack talked over his idea with his parents again and again. He argued his case so persistently and logically that he eventually won them over. Or maybe they knew that Jack would go ahead anyway. "Personally, I thought this project was the perfect fit for me—I was in search of an outlet for all my grief, and cancer was in need of a cure."

Both of Jack's parents had always supported his interest in science projects. Even when Jack and his brother blew out all the lights on their city block with a massive electricity project gone wrong, they continued their support. They made the boys promise to be more careful and let them continue to experiment. Gradually, reluctantly,

Jack's parents decided to support him in this latest quest. This was something Jack needed to do.

Google and Wikipedia: A Teenager's Two Best Friends

The support of Jack's parents was a major step, but now what? Jack realized that he needed to know more about pancreatic cancer, and he wasn't so sure he knew exactly what the pancreas was, anyway. He googled it. This was the beginning of a long program of research using internet search results, Wikipedia, and then journal articles.

As Jack read, researched, and identified all the questions he needed to answer, he realized how big a job finding a cure for pancreatic cancer was. Skilled medical researchers had been working on it for years.

As Jack continued to learn more about pancreatic cancer, he kept coming back to the fact that most pancreatic cancers are found too late, after the cancer has caused too much damage to the body. This happened to Uncle Ted. Jack remembered hearing Uncle Ted's doctors saying, *If only we caught it earlier.*

Pancreatic Cancer

Pancreatic cancer is both deadly and hard to detect. Apple cofounder Steve Jobs and astronaut Sally Ride died of this cancer.

The pancreas rests in the abdomen behind the stomach. This organ is about 6 inches (15 cm) long and shaped like a flat letter J. The pancreas has two main jobs. It makes chemicals that help the body digest food, and it produces insulin, a chemical that regulates sugar in the bloodstream. If your body produces too little insulin, you may get diabetes. People with this disease have too much sugar in their blood. This can damage their eyes and other organs and increase the risk of a heart attack.

Cancers begin when some cells mutate and multiply explosively. They stop following their normal programming and stop responding to the body's control system. Normally, the body's immune system kills these cancer cells right away. But sometimes the body's defense system fails. When this happens, the cancer cells grow and multiply unchecked, forming a tumor. At some point, the growth of these cancer cells interferes with the organs of the body. Doctors name cancers for the organ where they first form. When tumors get larger, they may metastasize, or shed small bits that spread through the body.

Pancreatic cancer is difficult to find because the pancreas is so deep inside the body. Doctors can't see or feel any tumors during a physical checkup. Most people do not show any early signs of having this disease. This cancer is so deadly because doctors rarely find it before it has spread throughout the body.

If detected early, when the cancer is small and still restricted to one organ, it is much easier to treat. Doctors can use some combination of surgery, chemotherapy, and radiation therapy to cut out, poison, or zap the cancer. These treatments work well for early-stage cancers, aiming to kill all or most of the cancer cells while killing as few healthy tissue cells as possible. Then the body's immune system can take over again. Once the cancer has spread, the standard cancer treatments don't work very well. The tumor breaks up into tiny pieces, and each small piece forms new cancers. Developing early detection methods, such as Jack's, is crucial.

Doctors can treat pancreatic cancer in the early stages of the disease. Jack realized that an early-stage diagnostic test could save many lives. Jack decided he had a new mission. He would discover a method for early detection of pancreatic cancer.

With this new goal, Jack decided his ideal test needed to be fast, cheap, and simple. He also knew that the test needed to be sensitive enough to detect the early signs of cancer. An ideal test would also be very specific to pancreatic cancer and not give false alarms.

To figure out what his detection test would look for, he started hunting for an early warning sign of pancreatic cancer.

Continuing his internet research, Jack came across the Public Library of Science (PLoS), an open access, online journal. PLoS contained a journal article that listed a database of different proteins found in people with pancreatic cancer.

Jack knew about proteins and their role in the structure, function, and regulation of the body's tissues and organs from his middle school biology class. He also had learned from his research that certain proteins can be indicators of disease. Often the indicator proteins show up in the early stages of the disease's progression. Jack says, "One little protein could be the key to detecting pancreatic cancer early, before it spread to other parts of the body and while it was still treatable."

Jack began combing this database to find the biomarker he was looking for. This database hunt was a daunting task since it contained eight thousand proteins. One by one, he searched the database and matched each protein against his set of criteria for the right biomarker. This matching involved hunting for the necessary information in other articles, since the database was just a listing. "I had no idea if I would succeed. But one thing was certain: my work had just begun."

Jack spent that whole summer researching each protein. Sometimes his analysis would take only a few minutes when the right studies were available online. Other times, the articles were not readily available or were inaccessible to the public. At these times, Jack had to buy the articles to finish his analysis of a single protein, and a typical article might cost thirty-five dollars.

These journal articles are written for academics and scientific researchers, so they are dense with technical terms. At first, Jack sometimes struggled with reading an article, reading it many times or taking several hours to work through it. He says it was not uncommon for one paragraph to take him half an hour to read. *Why were these articles so difficult to read?* he wondered more than once.

He kept his online dictionary open and looked up words he didn't know, which at first was a lot. Then he reread sections and thought about the ideas until they became clearer and then clear. "I kept moving forward . . . until eventually more and more words began to make sense to me." As his understanding of the fundamental ideas grew and as he learned the lingo of medical research, his research went faster.

After many months, Jack narrowed the list of approximately eight thousand proteins down to about fifty. This represented great progress, but those fifty proteins would be the toughest to evaluate because there was so little research on them.

Finally, nearing desperation, Jack struck pay dirt when one protein passed all of his tests. "That's it! Mesothelin!"

Excitedly, he told his mother. Thrilled, she asked him if that meant he had found the test, and he told her no, not yet. But he knew he was a lot closer.

Finding the biomarker he was looking for was a big piece of the puzzle. Mesothelin appeared in high concentrations in the blood of people with pancreatic, ovarian, and lung cancer from the earliest stages of these diseases, according to his research. The question became, Could he detect this protein in people? Safely? Easily? Cheaply? Quickly? Answering one big question opened up his next challenges.

Jack went back to his research and read all he could about mesothelin and how to detect it. Day by day, he kept reading and thinking. He even took scientific journal articles with him to school and read them during class.

Then, one day during biology class, Jack was reading an article from a science magazine under his desk and half listening to his teacher. The article described an unusual form of carbon called nanotubes (see page 45 for more information).

Nanotubes look like chicken wire wrapped to form a cylinder. Each tiny cylinder is just one atom thick and thousands of times narrower than a human hair. These nanotubes are very strong and conduct electricity well. *How cool!* Jack thought.

Meanwhile, the teacher was telling the class about antibodies. Antibodies are chemicals in your bloodstream. They are part of your body's immune system, fighting bacteria and viruses to keep you healthy. Our bodies make a specific antibody for fighting each type of invading bacteria or virus. The antibody binds to the invader until the immune system sends other cells in for the kill. The combined size of the antibody with its "captive" invader is bigger than either the antibody or the intruder.

Wait a minute! Jack thought. The body produces mesothelin in the early stages of pancreatic cancer. So, the body knows the cancer is there. There had to be a way to find out when the body was producing mesothelin. "And it was then, sitting in class, that it suddenly hit me: I could combine what I was reading about—carbon nanotubes—with what I was supposed to be thinking about—antibodies."

There is an antibody for mesothelin. When this antibody locates a molecule of mesothelin, it grabs onto it, forming an even bigger molecule. This bigger molecule, called an immunocomplex, would indicate cancer. The trick, then, was figuring out how to tell if the mesothelin immunocomplexes were in the bloodstream.

What if you stuck mesothelin antibodies into a mesh of carbon nanotubes? Jack compares this to meatballs in a pile of spaghetti. Embedding mesothelin antibodies in nanotubes creates a mesh that would react with only one specific protein, the mesothelin biomarker. "I was having one of those moments when it all began coming together in my mind. I could take these nanotubes and mix them with antibodies."

Then, if you put a drop of blood containing mesothelin on the nanotubes studded with antibodies, the antibodies would grab onto the mesothelin and form mesothelin immunocomplexes. The antibody with the mesothelin attached to it would stretch the mesh of the nanotubes.

Next, he needed to tell the difference between a bunch of nanotubes with antibodies and a bunch of nanotubes with immunocomplexes.

Carbon nanotubes conduct electricity and have a certain electrical resistance. This was one of the cool properties of nanotubes Jack was reading about. He figured out that because of the nanotubes' electrical properties, the stretched nanotube mesh with the immunocomplex would conduct electricity differently than the one with just antibodies.

When the mesothelin immunocomplexes stretch out the nanotube mesh, the electrical resistance increases because the conducting pathways for the electrons are longer and thinner. Both of these factors increase the resistance for the electrons as they flow through the nanotube network. Jack realized that he could measure this increase in electrical resistance. "I could feel the simple pleasure of all these puzzle pieces linking together in my head." This was the moment of breakthrough Jack had been working toward.

At home that afternoon, Jack realized that his network of nanotubes and antibodies would need a supporting structure, as nanotubes are

This illustration shows one possible carbon nanotube structure. The extraordinary electrical properties of nanotubes make them ideal for sensors, semiconductors, displays, photovoltaics, and energy conversion devices.

very delicate. He wanted a material that would be cheap, easy to work with, and provide the needed support. *Paper!* He realized almost immediately that strips of blotting paper would be perfect.

The whole concept was coming together. By combining these two ideas, antibodies and electrical resistance, Jack had found a way to test for pancreatic cancer. This linked with his discovery of the mesothelin biomarker from the medical database.

By mixing nanotubes and antibodies in water, dipping paper strips into the mixture, and then drying out the strips, he could make a cheap, easy to use pancreatic cancer detection device. In his first TED Talk, Jack compares this process to making chocolate chip cookies. "This should be easy," Jack thought. Well . . . not quite, as he would soon find out.

Electrical Resistance

Electricity is the flow of electrons through a conductor, such as a wire. Electrical current is the number of these small charged particles that pass a certain point each second. The amount of current in a circuit depends in part on the amount of force pushing the electrons through the circuit. The greater the push, the greater the flow of electrons, and so the greater the current.

Electrical charge moves more easily through materials with more free electrons, which are electrons not attached to an atom or molecule. Some materials, such as copper, are made of atoms that have many freely circulating electrons that are easy to push through a wire. When electricity moves easily through such a material, we say the material is a good conductor of electricity.

Metals are good conductors of electricity, and carbon nanotubes are too. Normally, carbon isn't as good a conductor of electricity as metals, but the structure of the nanotubes leaves free electrons.

A force, called voltage, pushes the electrons through a circuit. This push comes from either a generator, which uses magnets, or a battery, which uses chemical reactions. (For more on electrical motors and generators, see chapter 3.) Increasing the voltage in an electrical circuit is like turning up the water pressure in a garden hose—more current flows.

One other factor, called resistance, affects the amount of current flowing in a circuit. Sometimes electrons can flow easily though the wire, and sometimes fewer electrons make it past a certain point in a given time. If the voltage (pressure) has stayed the same, yet fewer electrons are moving through the circuit, then the resistance has increased.

Four factors influence resistance in a circuit. First, there is the type of material the current is flowing through. Second is the temperature of the wire. For most materials, such as copper, the hotter the wire, the slower the current travels. The length of the wire and the diameter of the wire are the last two factors affecting resistance. Shorter, thicker wires make it easier for electrons to flow than longer, thinner wires. Lengthening the wire increases the time for a given number of electrons to pass a certain point, as long as the voltage stays the same.

In Jack's test, when the immunocomplexes stretch the nanotube structure, the pathways the electrons can travel become both longer and thinner. This change in resistance is measured by an ohmmeter.

This relationship among current, voltage, and resistance is expressed as a formula. The amount of current (I) equals the voltage (V) multiplied by the resistance (R) or I = VR. The German physicist Georg Ohm discovered this principle in 1825. Most of the inventors in this book used this idea, called Ohm's law.

With his idea a clear picture in his mind, all he needed was a lab to test his plan and work out the details. He knew his mom would not allow him to do cancer research in her kitchen or anywhere else in their home. He needed a real lab and the right equipment for his plan to work.

With some more internet research, he found that he could write a proposal for his idea and then ask medical researchers specializing in pancreatic cancer for lab space.

Jack designed an experiment to test his theory. He then spent four months writing his thirty-page proposal. He explained how the mesh of nanotubes would snare the mesothelin antibodies and then measure any change in electrical resistance.

With his proposal finished, Jack went back to the internet to search for all the medical researchers at nearby universities who might help him. He sent off his proposal to two hundred professors at places like Johns Hopkins University and the National Institutes of Health. Naively, Jack assumed that cancer researchers would be eager to offer him lab space and assistance. *Who can refuse a kid?* he thought.

Then he sat back to wait. And wait. And wait.

Then the rejections began to come in. Most were polite and formal. Some were needlessly cruel. After three weeks he had over a hundred rejections and not one acceptance.

His parents sat him down to talk to him about being realistic. After all, if so many of the nation's brightest pancreatic cancer researchers did not think it would work, then it probably wouldn't. Maybe it was time to set aside this obsessive project of his that was taking so much of his time and energy. Maybe it was time to move on.

Everything Jack's parents said was practical and reasonable. But Jack thought maybe the world wasn't changed by being "realistic" or "practical." "My parents had a lot of arguments. All I had were three little words. 'But it works.'"

They continued to talk until they understood the depth of Jack's commitment. Giving each other a long look, his parents agreed to continue supporting his project. And perhaps most important to Jack, his older brother, Luke, looked over his research and declared that his plan would work. Though they competed fiercely at times, Jack greatly respected Luke's skill and success in science. "From that moment forward, I didn't doubt the validity of my idea again."

A month after his mass mailing, Jack had 192 rejections out of the 200 he had sent out. But one May afternoon, after coming home from school, he noticed a letter from Dr. Anirban Maitra of Johns Hopkins. "This is a really interesting proposal. Come in and talk about it," the letter said. Maitra was one of the most preeminent pancreatic cancer researchers in the world. Jack was elated. He scheduled to meet with Maitra right away.

On the day of the meeting, Jack was nervous, but his mother told him to relax and just be himself. Maitra met him in his office and put Jack at ease, listening patiently as Jack explained his idea. Then Maitra brought Jack to a small conference room and in came his researchers.

The researchers began to quiz Jack about his experiment. They peppered him with detailed questions for two hours. "They were relentless. It was exhausting. . . . Finally it was over. I had survived."

Looking around the room, Jack saw that everyone seemed pleased. Finally, Maitra was satisfied and agreed to give Jack lab space, access to the equipment he needed, and the help of a graduate student to learn his way around the lab.

Once he got started in the lab, Jack realized just how much he had to learn. A sudden sneeze contaminated many hours of work. He accidentally baked cell cultures in an incubator by leaving them in too long or at too high a temperature. Maybe most devastating was tripping over his own unlaced sneaker and sending his whole batch of test cancer cells crashing to the floor, ruined. Those cells had taken Jack two months to grow. "Now I would have to start from scratch," he recalls of the incident.

Jack also needed to learn to use the lab equipment to separate the antibodies he needed and then to merge the antibodies with the nanotubes. He learned to use a scanning electron microscope to find the right amount of antibody-studded nanotube mesh to coat strips of filter paper, which then became his test strips. He learned to graph out the results of the electrical resistance tests. It was hard work. Learning to use this equipment took Jack a long time, and he made frustratingly slow progress. He had many discouraging days, and at times he despaired that he'd never get it right.

After these mishaps, Jack felt embarrassed in front of the graduate students, who had developed their research skills over years. At break times when everyone gathered at a table, he felt he didn't have much to offer when the conversation turned to spouses and kids. He took to eating alone in the stairwell.

As Jack continued working, he found more and more problems he needed to deal with to make his idea work. Lonely and disheartened, progress seemed so slow to him. One dark and discouraging night, he went home and sought inspiration from a story about the life of famous inventor Thomas Edison, in which Edison got right back to work after a fire destroyed his factory. "I am sixty-seven, but I'm not too old to make a fresh start," Edison said.

> I made a conscious effort to see the setbacks as opportunities.
>
> —*Jack Andraka*

Jack tried to follow Edison's example and see what he could learn from his mistakes rather than becoming discouraged. Thinking about his own setbacks, he realized that many of them were because of either inexperience or the need to pay more attention to the details of his work. "I made a conscious effort to see the setbacks as opportunities and to remind myself that within each mistake was a clue that could bring me another step closer to creating an early detection method for pancreatic cancer."

The words of Uncle Ted came back to him: "Just slow down, Jack, you're going to be okay. Everything will work out."

Jack redoubled his efforts, working longer hours and working more painstakingly. He began going to the lab every night after school and past midnight on Saturdays. Jack worked through meals, instead taking brief breaks to snack outside the lab only to jump right back in. On Thanksgiving and Christmas, he was in the lab working. When he was too tired, he took short power naps underneath a stairwell. "I thought it was a great hiding space until, one time, I woke up from a nap and saw Dr. Maitra staring down at me with a look of complete confusion." For his birthday, he decorated his workspace with party hats and streamers and kept on working.

One night in December, after seven months of work, was particularly difficult for Jack. He had made mistakes on three different batches that evening. Each time, he needed to begin the whole long process all over. Each test run could take up to three hours to complete. He was thinking of quitting for the evening but decided to do one more trial.

On this fourth trial of the long day, Jack worked painstakingly on the familiar and tedious process. Finally, hooking up his treated test strips to the ohmmeter, he saw something different. *Wait! What was this?* he thought. "The numbers showed that my test with little paper strips had detected the biomarker!" It worked! Jack quickly checked his math again. The thought flashed in his mind that maybe because it was so late at night after such a long day, he was just seeing what he wanted to see. "I ran back over and checked my results again. My hands were shaking as I held the ohmmeter. There it was—my hypothesis was correct." Jack's experiment had shown a direct relationship between the amount of mesothelin on a test strip and the amount of change in the electrical resistance.

Jack was ecstatic and looked up to see who he could share the good news with. It was 2:30 a.m. on a Sunday morning, and everyone else in the lab was long gone. But Jack's mother was there, asleep in the car, waiting. She'd been waiting for hours that night, as she had been doing every lab night over the past seven months. Walking up to the car, Jack called out to her. "'Hey, Mom, guess what? . . . It's working!' I screamed at her. . . . She began screaming. I was screaming. We were both screaming."

Together, they celebrated and screamed their excitement all the way home in the car. Only one thought made Jack sad at that moment. "The one person I wanted to share my joy with more than anyone else was Uncle Ted. . . . He would have loved this moment most of all."

The next day, Jack, wondering if he had dreamed it all, double-checked his calculations. Then he emailed his results to Maitra, who sent back an enthusiastic reply and congratulations.

Next, they tested Jack's process on real human tumor cells. A month later, the results were in. This step worked too.

How was Jack going to let others know about his findings? These results were more than worthy of an Intel International Science and Engineering Fair (ISEF) entry. And Jack had great success at science fairs. "I'd been dreaming about that competition for years."

Success and ISEF

Jack wondered if a good showing at ISEF would help get his pancreatic cancer test out into the world and save lives. What if this test had been available for Uncle Ted? Would Uncle Ted still be alive? If getting his test noticed could help people like Uncle Ted, then it was worth doing, Jack decided.

Jack had just four months to get ready. Although he had entered and won many science fairs since grade school, this was the big time, the ultimate in science fairs. As he put it, the ISEF is "like the Super Bowl, World Series, NBA Finals, Stanley Cup, and Olympics all rolled into one and multiplied by a factor of three." Using his pancreatic cancer test, Jack entered science fairs that qualified him for ISEF and gave him the practice he wanted. Soon he was on this way to Pittsburgh, Pennsylvania, and ISEF.

At ISEF, Jack met young scientists from all over the world. These teens shared their experience and their innovative ideas. The eighteen hundred participants got to ISEF by winning in local, regional, state, and national competitions. Everyone here had brains, determination, and science chops. The competition was fierce.

The whole ISEF experience is amazing. Bright, young scientists come from all over the world with exciting ideas. Venture capitalists and executives from major technology companies study the vivid displays. World-renowned professors from elite universities come to meet the entrants. Nobel laureates and science geeks crowd the aisles. Everyone is in search of new ideas and new talent.

Jack reveled in the excitement of the fair and in meeting his clan. As he roamed the aisles, looking at the other exhibits, he was awed by the work of his peers. Other finalists had discovered new proteins or contributed to preventing Alzheimer's disease. *How can I compete with projects like these?* he wondered.

With his characteristic optimism, he set about preparing himself for the judging day. He struck up a friendship with two finalists from New Jersey, and they agreed to go to the dance party together. Jack loves partying and dancing, so he reveled in letting off some steam. It was very late (or very early) when he made his way back to his hotel room.

He awoke to a pounding on his door. He stumbled out of bed to find Valerie, his fair coordinator, imploring him to get going or he'd be late for the judging. Only he couldn't speak—he'd lost his voice. She was back in a flash with bottles of Gatorade and demanded he drink.

With Valerie's prodding, Jack dashed to the convention floor. Once in his booth, Jack fell into his practiced routine. He had prepared and practiced his pitch many, many times. He enjoyed

The ISEF is the world's largest student science competition. Thousands of people flock to the competition, which takes place in a different US city each year.

meeting the people who came to see his exhibit and liked the challenge of answering the questions of the judges. He wasn't nervous. He was in the zone. "I absolutely relished the opportunity to talk about my project," Jack says. His pitch and project attracted a good crowd around his booth, and he noticed several judges viewing his project. He knew this was a good sign.

Then the judging was over, and the wait began. At the special award ceremony, Jack was thrilled to win six prizes. His family and friends were happy for him. Even if he won nothing else, his project and his effort were recognized as significant by the judges and his peers. When he won the Best in Category award, he realized "that I was now in the running to compete for the biggest award of all—the Gordon E. Moore Award"—the grand prize. The Moore Award is for both the best research and the most potential for impact on the world.

At the final ceremony the next day, Jack and everyone waited as the other winners were announced. Then, finally, the judges announced the winner of the Moore Award: Jack. *That's me!* he realized. He leapt from his seat. Not one to walk to the stage quietly, he gasped for air as he ran, screamed, and lifted the presenter off the floor in a big hug. "I had to remind myself to breathe," Jack remembers.

As he stood onstage, the memories flooded in: the long hours in the lab, the bullying he endured at school, Maitra of Johns Hopkins who gave him a chance, his family who supported him and, of course, Uncle Ted. "The weight of it all was almost too much to bear."

Aftermath

Jack celebrated with his family, who had been there supporting him all along. Many well-wishers congratulated him on this enormous accomplishment. One of them was a stern-faced, middle-aged man, who grabbed his hand and thanked Jack. Jack asked why. The man said that six years ago, he had lost the love of his life to pancreatic cancer. With tears rolling down his cheeks, he said, "Looking at you, watching you talk . . . it makes me feel hopeful again." Jack told this stranger about Uncle Ted as they hugged. " 'I'm sure he is proud of you,' he [the man] said as he gave my hand one last squeeze before walking away." Later, Jack recalled that of all the adulation he received, including the Moore Award, the kind words of that stranger meant the most to him.

Media Attention

On the heels of his win at ISEF, Jack found himself involved in a whirlwind of publicity. Within a week he was doing an interview for CNN's *Early Start* with Alina Cho. Thousands of requests for interviews poured in. Morley Safer of *60 Minutes* even came to Jack's house to interview him. Bill and Hillary Clinton invited Jack to an event sponsored by their Clinton Global Initiative program. He and his mother got to meet the pope. A very special invitation came from Michelle Obama to be her guest at the State of the Union address. Jack got to speak with President Barack Obama afterward.

Mail from well-wishers and fans soon flooded Jack's mailbox. Many writers shared stories of losing loved ones. The most poignant urgently requested information on where to buy Jack's test for themselves or a family member. How soon could they get it? Where could they get it?

In truth, Jack's pancreatic cancer test would take years to get to market, and by 2021, the process still wasn't complete. While his work is truly remarkable, more research, more testing, and a long approval process are needed before doctors can use it.

And he got a lot of mail from haters and homophobes targeting him for being openly gay. Jack says, "I never really set out to become a gay role model. . . . I wanted to share my ideas and become a scientist. But being gay is part of who I am, so when it came up in an interview, I decided to be honest." Jack remembers going to science fairs and participating in other science events and wondering why he didn't see more LGBTQ+ people. He decided that "maybe my story would make it easier for the next kid who wanted to come out."

While most of his publicity was very positive, Jack has received some critique of his research by other scientists. One measure of the importance of Jack's work is the level of scientific scrutiny it is now receiving.

"I really appreciate their concerns," he says, "because their concerns made my project better. It was definitely a learning curve."

School at Stanford

Jack graduated from Stanford University in 2019, where he majored in electrical engineering and anthropology and continued his work with disease-detecting biosensors. As a junior he won a Truman Scholarship, which is awarded to just fifty-nine students across the country who want to pursue graduate studies leading to public service.

Jack entered medical school at Stanford and also went on to study for a master's degree in public health, specializing in global health issues. "The [medical] degree would enable me to frame my engineering background within medicine and provide me with an appreciation for the clinical realities of working in global health."

These two degrees tie together his long-term interests in using science and engineering to improve health outcomes. He plans to become a public health physician focusing on the vast inequities in health care around the world.

Jack also makes time to advocate for math and science, especially for students who have never had an interest in science, technology, engineering, and mathematics (STEM) subjects. Jack says he has lots of ideas for transforming schools. He believes that rather than memorizing facts from textbooks, students should be making things at shops or experimenting in labs.

Jack continues to work with pancreatic cancer groups and speak at conferences. Outside of his studies and his research projects, Jack finds time to hike and climb with his friends. An avid kayaker, he competes in whitewater contests.

Jack advises kids and their parents to explore, experiment, and give free rein to their creativity and imagination. "Make sure to be passionate about whatever it is you get into, because otherwise you won't put the right amount of work into it. . . . No one will be excited about your work if you're not excited about it."

2

GITANJALI RAO AND TETHYS, HER LEAD DETECTION SENSOR

"This is Opemipo. He's among the thousands of children and adults exposed to lead in drinking water in Flint, Michigan," says a twelve-year-old Gitanjali Rao. She's explaining her invention, a device to detect lead in drinking water, to a room full of scientists, engineers, and executives at the 3M headquarters in Saint Paul, Minnesota. Because lead dissolved in drinking water is very toxic, her invention could prevent serious brain damage and save lives. Its potential, and the remarkable science behind it, brought her to this stage. After five months of preparation, she was in the 2017 finals of the Discovery Education 3M Young Scientist Challenge, the nation's premier science competition for middle school students. Gitanjali just had to convince this room of judges about the value of her idea and her work.

Triggering Event: The Flint Water Crisis

Gitanjali first heard about the hazards of lead in drinking water in the news when she was ten years old. "Anything we flipped on, we saw something about the water crisis in Flint," she says. "It was something that people just kept talking about but weren't doing anything about it."

Along with the constant buzz in the news, she learned more about the impact of lead contamination while attending a STEM club meeting at her local chapter of 4-H, an education and professional development organization. The 4-H leader sparked a discussion about water pollution, introduced a hands-on activity removing oil from water using feathers, and emphasized the tragedy in Flint.

Gitanjali heard how the city of Flint, Michigan, had switched from using clean water from Lake Huron to the polluted water of the Flint River to save money. The city planned to clean up the water before piping it into people's homes, but they failed to treat the water to make it less corrosive, as their previous supplier did. So, the inadequately treated water corroded the old lead pipes in the city and poisoned the drinking water with lead. CNN reported that both the Environmental Protection Agency (EPA) and Virginia Tech researchers found dangerous levels of lead in the home drinking water of Flint residents.

At first, the city claimed the water was fine. Later, they faked the water quality test data. Eventually, people got sick and doctors detected high levels of lead in their bloodstreams. Residents had been drinking toxic water and didn't know it. "It's really unfair that kids don't have clean water to drink, especially because water is a basic right that everyone should have," Gitanjali asserts. "I decided to take action."

Lead is bad for everyone and can cause a range of illnesses, damage to the nervous system, and injury to internal organs. Lead is especially dangerous for young children and can cause permanent brain damage. Scientists unaffiliated with the city came to Flint and tested the water, showing alarming levels of lead. President Obama declared a state of emergency, and officials advised people to drink only the bottled water that was trucked into the city in massive quantities. It horrified Gitanjali and her friends to hear about this crisis and the danger to children. Gitanjali knew she wanted to do something about this disaster. "It was appalling to me the number of people who got affected by lead in our water."

This photo illustrates, from left to right, a lead pipe, a corroded lead pipe, and a lead pipe that has been treated with a protective orthophosphate coating. Many cities use orthophosphate coatings to reduce corrosion while they work to replace sources of lead contamination.

Gitanjali's research showed that lead contamination of drinking water is more widespread than most people believe. She found that about fifty-three hundred water systems across the US have unsafe levels of lead or copper. Could all these communities have contaminated drinking water?

Gitanjali remembers worrying about the drinking water in her home. Was it contaminated? Were they poisoning themselves without knowing it? Gitanjali and her parents decided to find out and bought a test kit.

Gitanjali watched her parents as they struggled with performing the test. The kit contained chemically treated strips of paper. These strips change color when dipped in a water sample. The amount of color change is supposed to show the amount of lead. The problem is that only very slight differences in color mark the difference between safe water and dangerous water.

Interpreting these slight color differences can be tricky at best. How much of a color change was safe? Many people, including Gitanjali and her parents, feel unsure about their ability to accurately interpret results of these home test kits. For toxic chemicals like lead, Gitanjali wanted more understandable water test results and felt that families shouldn't take life-threatening chances based on ambiguous data.

Gitanjali knew that drinking water would ideally contain no lead and that the EPA had set an "action level" of 15 parts per billion (ppb). When water tests above this level, the federal government requires water suppliers to take action to protect their communities. This EPA limit recognized the practical reality that lead could contaminate drinking water in many ways, including via the lead pipes in the houses of many people and that some amount of lead was almost inevitable. But there is a vast difference between trace amounts of lead (1 to 2 ppb) and the 13,200 ppb found in some Flint homes. Gitanjali noted that most test strips just indicate whether lead is present or not, but not the amount. She wanted a test to show if the amount of lead was in the range the EPA considered allowable.

The next best solution is to send off a water sample to a professional lab for testing. These tests take a couple of weeks, and householders need to follow strict rules for collecting the water sample. The cost for water testing at commercial labs is expensive: fifty dollars or more per sample. Another option, having a water quality technician come to your home, costs even more. After watching her mother follow the exacting procedures to get an acceptable water sample to send to a lab, Gitanjali knew there must be a better way. "I wanted to do something to fix this, so it not only helped my parents but also Flint, Michigan, and places like this around the world."

MIT Nanoparticle-Sensor Article

Gitanjali had long been an inventor and tinkerer. The gift of a chemistry set when she was six or seven years old fired her imagination. Mixing the chemicals and watching the reactions was like magic to her. One of her first inventions, created when she was in elementary school, was a tool that detects snakebite severity by identifying the type of venom in the bite.

When bitten by a snake, a person's body brings more blood to the bite area and raises the temperature. Gitanjali found that different types of venom have distinct heat signatures when viewed through a thermographic camera. This invention, called Asclepius, became her first entry in the 3M contest when she was in fifth grade. She often names her inventions after Greek gods because of her fascination with mythological characters. Gitanjali still has her prototype of Asclepius, "my first time working with a laser cutter," on her desk in the room at home set aside for her science projects.

Ever since those first forays into science, Gitanjali has been a fan of the *MIT Technology Review*, a magazine published by the Massachusetts Institute of Technology (MIT), which helps her keep up on the newest ideas and inventions. Not long after concluding that water testing for lead needed improving, she read an article on the *MIT Technology Review* web page about using carbon nanotubes to make sensors to detect hazardous gases in air. By running an electric current through the nanotubes, the MIT scientists could identify the type of gas particles trapped in the mesh.

This was a eureka moment for Gitanjali. Excitedly, she realized this same principle could apply to questionable drinking water. She immediately saw how she could build a sensor that would trap any lead particles in drinking water in the nanotube mesh. With this vision of a device to detect lead in drinking water clear in her mind, she set out to

Carbon Nanotubes

The MIT article Gitanjali read explained how the device used carbon nanotubes to trap gas particles. Nanotubes are small carbon molecules that look like nylon mesh cylinders when viewed through a powerful microscope. Each intersection in the mesh is a single carbon atom. Gas particles too big to pass through get trapped in the netting.

Some materials conduct electricity more than others. Scientists say these materials have lower electrical resistance. Gas molecules attached to the mesh change the resistance of the carbon nanotubes. By running an electric current through the nanotubes before and after adding a sample of gas, the MIT scientists could measure the changes in resistance caused by the trapped gas particles. These changes allow the scientists to identify the trapped gas molecules.

build it. She named it Tethys, after the Greek god of fresh water. "My solution uses carbon nanotubes to detect lead in water faster than any other current techniques. It has a carbon nanotube sensor, to which special atoms are added that react to lead," she says of Tethys.

How long does it take to develop her inventions? "Usually when I'm coming up with an idea, the formulating takes the most amount of time. When I'm doing all the experimenting, doing all the designing—that happens quickly because it's the part that I enjoy the most." Like many scientists and inventors, Gitanjali first wrote her ideas and questions and made her drawings in her notebook. With Tethys, the incubation period took a couple of years, but she kept coming back to it, as she knew it was a problem she wanted to solve. She just couldn't forget the people of Flint, Michigan, because it seemed so unfair for the community to suffer through this. She kept researching and thinking to figure out what she could invent and how she would do it.

"My big break was when I heard about the MIT article . . . and that's when I started to connect it back." She recalls that "from the MIT article, I started working."

3M Application Deadline

Gitanjali realized that a detector for lead in water would make a significant project for the 3M contest. And it would be a way for her to present the concept to the scientists and engineers acting as judges. Through the contest, she could "see what people think about it." Gitanjali says, "I love feedback . . . because that lets me improve."

But time was short. The deadline for entry was only a few weeks away. There was so much to do! Gitanjali knew she had to turn a good idea into a well-thought-out prototype and complete her entry quickly. The entry required a one- to two-minute video explaining the idea and the science as part of the application. This was the first time she'd ever put together a polished video. She taught herself how to use a video editor, which was a struggle for her. But she stayed calm. "I don't really pressure myself. I'm very competitive, but I don't pressure myself that much because I always know it's a learning opportunity."

Gitanjali set to work on her prototype and video. She knew what she wanted it to look like from the outset. Using a basic idea similar to the sensor in the *MIT Technology Review* article, she began prototyping her concept with craft materials she had at home. "It was pretty ugly, to be honest," she says of her first efforts.

Were there failures and dead ends along the path for Gitanjali? "Oh, gosh yes, way too many. Way, way too many, to the point where I can't even say every single one of them. There were so many barriers that I faced during the entire process. Going into Tethys, I was terrified. I knew things weren't going to work the first shot and I didn't know what I was going to do if didn't work. So, it was a challenge," she says with a small smile, "but I evidently got it done."

As her father says of Gitanjali, "It turned out she had a lot more determination."

Gitanjali's application video shows that she had fully planned how her device would work and how its three major components—a sensor, a multimeter, and a smartphone app—operated. The sensor would chemically react to any lead in a water sample. Any chemical change would slow the flow of electricity through the sensor in proportion to the amount of lead in the water. The multimeter, a device used by engineers and electricians to measure many electrical properties such as voltage and current, measured any change in the speed of electrical flow. The smartphone app would then calculate the amount of lead in the water sample based on any slowing of the current (any increase in the resistance of the circuit) and display the calculation.

Her video simulated how Tethys would work. The prototype in her video was a cardboard box. "I think it was an iPhone box that I painted over. One side of the box was hollow and there were wires sticking out of it. I think I destroyed an Arduino [board] because I superglued it onto the box." Gitanjali's application video shows how effectively a well-designed simulation can present a complex idea.

Gitanjali finished her prototype and her video just in time. She submitted her entry and waited for the results. During those two months, Gitanjali kept busy with school and all of her other activities. "To be

An eleven-year-old Gitanjali Rao demonstates how Tethys works to a science class in Colorado.

honest, I forgot about it in the middle of that time," she says. "But once I got the result, I was beyond excited."

Good News

In a few weeks, Gitanjali got the news that she was a finalist in the 3M contest. When 3M first called, the family missed the call. "I was probably playing *Just Dance* or something. And that night, they called us again." *It's 3M*, her mother called out. "I ran to the phone and picked it up and basically screamed into the phone. I probably deafened the other person on the line," she recalls.

She was very excited, but she knew her work had only begun. The contestants had only three months to complete work on their final prototypes and prepare for the competition. Gitanjali had a colossal job ahead to make her invention ready for presentation to the judges. Once she had selected carbon nanotubes for the sensor, Arduino components to build the signal processor, and Bluetooth technology for a smartphone app, she set to work experimenting.

Each of the three main parts of Tethys would need significant work that summer. First, she needed to prove that the sensor would work. After that, she would make the multimeter and then program the smartphone app. She also made a plastic case for her device using a 3D printer she learned how to operate.

This was the first time she'd ever coded a smartphone app. She spent much of those three months creating the app. Developing Tethys was a time of many firsts for Gitanjali. She says it was also her first time coding an Arduino without help, as well as her first time working in a lab to run her tests.

Each of the finalists were paired with one of 3M's scientists as a mentor. Over that summer Gitanjali worked feverishly with her 3M mentor, Dr. Kathleen Shafer, by phone to get her device ready for the finals. "She's been so much fun to work with, and I have looked forward to her calls every week," Gitanjali says. "On the first Skype call, I was a bit nervous to talk to someone so knowledgeable and an accomplished scientist, but as soon as I started talking to her, she made me very comfortable and I wasn't nervous at all."

They talked about everything from the principles of scientific research to safety practices when working with a toxic substance such as lead. "[Shafer] led me through design thinking [a creative problem-solving process] and scientific thinking," says Gitanjali. "She merged them together. She helped me to slow down in some places [as well as pointed out] where I should speed up and where I should focus most of my time.

"Our talks started out being once every two weeks. Toward the end, it was four times a week. You're cramming for the last couple of days. It was a rule of the competition to have 3M products involved. I had four, maybe, in the beginning. By the last week, I had almost eighteen," she says. Toward the end, Shafer coached Gitanjali on her presentation for the finals.

Did she run into roadblocks along the way? "Oh, yeah. The biggest roadblock was the fact that I was eleven years old and trying to find a lab to do all my tests in. Most people thought of me as an eleven-year-old assigned to a project, but not going to take it seriously. I couldn't really blame them." Gitanjali, energetic and persuasive, convinced the chemistry teacher at a local high school to let her use lab space for her research.

Gitanjali thanks her 3M mentor for some valuable life lessons. "[Shafer] taught me to reach out and ask for help. Before, I would hesitate to ask a question to someone whom I haven't met before. But Dr. Shafer encouraged me to reach out to college professors and high school teachers for either space to perform my tests or to ask a question related to my research." Her parents also helped her get materials and equipment.

Although her workload was immense and experiments didn't always go as planned, Gitanjali persisted. "Never give up," she says. "There is no limit to the number of times that you could try an experiment or do it all over again. It might seem frustrating once or twice when your experiment doesn't go the way you want it to. Just take a deep breath and try to solve your way through it." Complimenting her 3M mentor, she adds, "I learned to be diligent and persistent from Dr. Shafer. She always listened to my failures and provided me alternate paths to keep moving ahead."

Building Tethys

The Sensor

For Gitanjali, the sensor was the most difficult part of the test to develop. Like Jack Andraka, she needed to find just the right amount of lead-sensitive chemical to use. To test a water sample, you dip the sensor in the water to see if it contains lead. The sensor has paper strips loaded with carbon nanotubes. This nanotube mesh contains a substance that reacts to any lead in the water and produces a chemical change. The more lead in the water, the more chemical change takes place. The more lead compound trapped in the nanotube mesh, the less current flows.

"It's like speed bumps for electrical current," Gitanjali explains. But when the cartridge is dipped in clean water, the flow of electricity doesn't change. Gitanjali ran many tests that summer to be sure this idea worked as she expected.

Gitanjali taught herself most of the chemistry she needed for Tethys. Her middle school chemistry class gave her a start, but she needed to go much deeper to complete her invention. The internet was a great portal for learning, and she found everything she needed online. She took online courses and read scientific articles. "I actually had to do a course on how to read scientific papers," she says. At first, it was difficult wading through the dense scientific articles, but it became easier as she learned the terminology. "It was a struggle, but it helped so much because now I can open up a scientific paper and read it all the way through and explain it," she says. "I just read by myself for hours on end every single day."

The Multimeter

In Tethys, the greater the electrical resistance the multimeter finds in a sample of water, the more lead there is. Gitanjali built her own special purpose multimeter for Tethys using Arduino parts.

The Arduino company makes microcontroller boards and kits that are inexpensive and easy to use so students can create sensors or devices to operate robots, games, or other electronic gadgets. "I've been messing around with Arduino since I was five, maybe," says Gitanjali. "But I'd never done anything that serious. But suddenly, I [had] to design an entire multimeter outfit."

The Smartphone App

The multimeter sends the information about the current to a smartphone via a Bluetooth wireless connection. Gitanjali wrote this smartphone app using a tool developed at MIT. While she had done some coding before, with the help of the tool, she wrote a complete app that summer.

The app calculates the amount of lead in the water sample. Then it displays a familiar stoplight image to show the test results. A green light shows the water is safe. A yellow light indicates caution, that some lead is present but at low levels.

"But when the cartridge is dipped in contaminated water, the lead in the water reacts to the atoms. This causes increased resistance in the electron flow measured by the Arduino processor. The more lead in the sample, the more the electrical current is slowed," Gitanjali explains. When you see a red light, "the water isn't safe to drink."

Gitanjali spent much of that summer working out these measurements. "Over the past couple of months, I have done so many experiments, and I have failed quite a few of them," she says. "But if I didn't make any of those mistakes, I wouldn't know this [my invention] could have been twenty times better," Gitanjali offers. "Through learning [from these failures], I finally got my experimentation and prototypes done," she says. "Then, everything kind of just came together, and I came to the [3M contest] with my complete prototype. That's the point when I knew I would be ready for this presentation."

One of Gitanjali's teachers, Simi Basu, says, "She's one of many who love science at school, but one of the few who turned an idea into an invention."

To Saint Paul for the Finals

Finally, after these long summer months, Gitanjali was ready for the big event where she would present Tethys to live judges. Along with all the other finalists, Gitanjali and her family traveled to Saint Paul for the deciding rounds of the competition. One of the first people to greet Gitanjali was her mentor, Kathleen Shafer. "Meeting my mentor was probably the highlight of the day. I loved seeing her since we communicated through online calls for the past three months."

That first day she also met the other contestants, all middle schoolers who shared her interests in science and technology. "I made some awesome friends there too. Their ideas were [really] amazing. We spent a good three days stressing out together. So, it was a lot of fun."

Following the introductions, Gitanjali and the other finalists toured the 3M labs and facilities. She could give free rein to her curiosity. "I got to ask a ton of questions and learned so much that I hadn't learned before.

"We got to do a lot of fun activities too. It was an awesome experience!" Along with the tours and the picture taking, Gitanjali and the other finalists were teamed up for additional rounds of the contest before the final presentation. The contestants had two on-site challenges to complete, which the judges scored and counted toward the final rating. Gitanjali worked well with her partner in each event and came up with clever ideas on the spot.

For the first challenge, Gitanjali and her partner needed to devise a new design for preventing food spoilage using a set of 3M materials the contest coordinators provided. Gitanjali says she thought this was very difficult, though between the two of them they had a workable solution when the buzzer sounded.

For the second challenge, new pairs of contestants had to build a Rube Goldberg machine that used different types of energy to activate various parts of the mechanism, such as a chemical reaction, a beam of light, and sound vibration. Planning and constructing this device while the clock was running was tough but exciting. Gitanjali says that contestants not only needed to be creative on a deadline but also meticulous so that all runways were aligned. She recalls with a laugh that she and her partner forgot to place a marble that would spring a mousetrap. Once they righted that miscue it worked admirably.

Originally invented by the cartoonist Rube Goldberg, Rube Goldberg machines are hilariously complicated devices that perform simple tasks, such as a twelve-step machine that cools a bowl of soup. People make Rube Goldberg machines for fun. Since Rube Goldberg machines often use simple machines, such as pulleys and levers, and illustrate mechanical principles, science educators often use them as teaching tools.

Between these activities, each contestant also continued to work with their mentor to prepare for the final event. Gitanjali and Shafer practiced Gitanjali's speech together and walked through her demonstration. Gitanjali shared and went through her slides while Shafer coached and gave her final advice. "I got to spend a lot of good quality time with her there," Gitanjali says.

The Big Day

At last the big day arrived. "I'm excited to talk in front of the judges, answer some of their questions, and let them know about my device so that this could spread awareness," Gitanjali said before giving her final presentation. Soon Gitanjali found herself in front of the judges and guests for her five-minute presentation and the five-minute question and answer session. "I was really stressed," she says. "Just from the second that I walked up on the stage, my hands were shaking."

While she may have felt stressed, her presentation was clear, persuasive, and fast-paced. Her passion and mastery of her subject were very convincing. She moved smoothly through her demonstration and explained Tethys effectively. At the end of her presentation to the judges, she fielded challenging questions from her audience of 3M scientists with assurance. Despite any nervousness, her audience saw a well-rehearsed, very poised, and very convincing young scientist.

"When I think of [Gitanjali], I see a young person who is tenacious and determined as a researcher, curious about the world around her, and possesses very sophisticated communication skills," Shafer said of her mentee.

And then it was over. When Gitanjali left the stage, she entered a small room with food and snacks for the presenters. Then she rejoined the other contestants at the back of the auditorium to hear the rest of the presentations and encourage her new friends. Gitanjali says they all cheered for one another and developed a strong sense of camaraderie. Everyone anxiously awaited the results, to be announced at the banquet that evening.

The company helped fill the time. After everyone finished presenting, all the contestants went to the greenroom to take down their posters and pack all their demonstration props for shipping back home. Games, snacks, and a lounge area were provided so they could chill out. After the contestants changed for the banquet, the guests and parents began filling the hall.

The banquet was a big affair. Video highlights played, and the previous year's winner gave a speech. Then, when they announced that Gitanjali had won, she was ecstatic. Celebrated by the audience, congratulated by other contestants, Gitanjali glowed. One judge later commented on Gitanjali's performance, "She blew us away!"

A 3M executive presented her with a check for $25,000, the top prize money. Gitanjali told the audience that she would use some money for developing Tethys into a commercial product; donate some to organizations she volunteers for, such as the Children's Kindness Network; and save some for college. She plans to study genetics or epidemiology at MIT. "I think that science can really make a difference," she says. Then there were more publicity photos, interviews, and congratulations from family, her 3M hosts, and her many new friends among the other contestants.

After the initial excitement died down, Gitanjali reflected on her experience. One of the best parts for Gitanjali was meeting the other finalists. "I've made lifelong friends here who enjoy STEM as much as I do," she says. "I'm so glad to have met all these wonderful people with me on this journey who experienced the same experiments and failures as me. When I came here, I was amazed. I thought, 'Wow. Their innovations are amazing,' " she says. "It's not just about there being one grand-prize winner. We all did a wonderful job and we're all winners."

After the 3M Contest

What will Gitanjali do next? First, she still has to finish high school. She plans to commercialize Tethys and make it available for families to test their own water safely and simply. She thinks Tethys might cost around twenty dollars when produced in quantity, making it affordable for homeowners and renters. Like her family, most people have little idea about possible contamination of their drinking water from lead. The EPA has identified fifty-three hundred water systems across the US with unsafe levels of lead.

Those systems serve around eighteen million people, according to CNN. Even if the water from the utility company is pure, the old lead piping in many houses may have begun to corrode and contaminate the water. Tethys's twenty-dollar price seems a bargain for peace of mind about the safety of the drinking water in your home. Gitanjali also thinks schools and hospitals should test their drinking water for lead.

While touring the facilities at Denver Water, the local water treatment plant, Gitanjali met Selene Hernandez-Ruiz, a lab manager. Hernandez-Ruiz invited Gitanjali back to use the facilities to continue developing Tethys. Excited by the discussion and the well-equipped labs, Gitanjali asked if she could come back every day. The two hit it off and started working together. "Right now, I'm looking at interference with other chemicals in water apart from lead," she says. "What if the carbon nanotubes accidentally bind to fluoride? So that's what I'm trying to tackle." Gitanjali appreciates the opportunity to work with Hernandez-Ruiz at Denver Water. "This gives me the potential to take [Tethys] out there. I know my device can be accurate."

Gitanjali plans to extend her basic design to test for other toxins dissolved in drinking water, such as copper or nitrates. A multitoxin test sensor would make Tethys even more useful.

Tethys might also help local municipalities and communities by "crowdsourcing" dissolved lead readings in an area. This data from community members could show patterns of lead contamination "hot spots" and launch community action to improve water quality.

Gitanjali leads a busy life. "I am a freshman in high school, so I'm really focused on school mainly. Apart from that, I have a whole bunch of interests. I play sports, I fence. Then I have also been working on a new device. I don't know if you've heard about it. It's a device that helps to diagnose opioid addiction at an early stage. I'm working on getting that out of the prototype stage and into something that can actually be used," she says.

Gitanjali is also active in DECA, an organization helping high school and college students develop real-world business skills. DECA sponsors regional and national challenges and competitions. She especially likes learning about entrepreneurship and says these skills will help her commercialize Tethys and her new opioid detection tool.

"One of our family friends became addicted to prescription opioids after a car accident. It's scary to think it can happen to you without you knowing it. I wanted to come up with a way to diagnose prescription opioid addiction at an early stage so you can take action earlier." Her tool, which employs artificial intelligence (AI), could help physicians by automating an early stage of the opioid addiction diagnosis. Epione, named after the Greek goddess of pain relief, examines the amount of additional protein produced by the bodies of those addicted to opioids. "It creates almost a spectrum of what a [protein] sample would look like without addiction and what a [protein] sample would [look] like with addiction," she explains.

"And then I'm working on an app called Kindly, which is incorporated in schools to prevent cyberbullying," says Gitanjali, describing one of her newest projects. "Kindly is basically like Grammarly for bullying. . . . It doesn't allow you to send [a message] if it has bullying in it." Kindly displays an alert and then builds in time for someone to take a moment to think about their message before sending or revising it. "It's definitely from a teenager to a teenager. That's really my goal for it."

Kindly is an outgrowth of Gitanjali's work with the Children's Kindness Network. This organization, founded in 1998, helps stop bullying on playgrounds, in schools, and in neighborhoods. "I've always been a firm believer that kindness is just something that you don't necessarily pick up. It's something that needs to be taught at a young age, so you continue that in the future," she says. Gitanjali had been a volunteer for a long time and is on the organization's national board.

In addition to winning the Discovery Education 3M Young Scientist Challenge, Gitanjali has also won the Paradigm Challenge, been selected as STEM Scout of the Year for her snakebite identification device, published a book about the world from her baby brother's point of view that she illustrated herself, reported for *Time for Kids*, won the International Aviation Art Contest four times, been a finalist for an engineering contest inventing tools for astronauts, been awarded a Davidson scholarship, given two TEDx Talks, and been named to *Forbes* magazine's "30 under 30" list for young innovators and leaders. In December 2020, *Time* magazine announced that Gitanjali was their first-ever Kid of the Year, selected from an impressive field of five thousand amazing young people. Gitanjali also plays the piano and the clarinet and has won acclaim for her performances. Her hobbies include singing, swimming, classical Indian dance, fencing, playing the bass guitar in a rock band, and baking. She's also getting her pilot's license. During the pandemic, she says she's done an incredible amount of baking.

How does she find time to do everything? "I think it's just time management," she replied. "It's just something that I picked up. I still manage to do all the things that I like, while also hanging out with my friends. It was hard in the beginning," she concedes, thinking of other things she has done and would like to do. "But I've really cut down a lot of the stuff going into high school this year. It's just about knowing that what you're doing is stuff that you like to do and then being about to push it all into one day. That sounds really hard and really bad, but it's actually not as hard as you think it is." Also, she says, "My parents kept me off social media."

She credits the immense support and encouragement from her parents for her many successes. "My dad introduces me to new technologies he comes across or sends me resources if he reads about something new. He sometimes simplifies the concept in such a way that it is easy for me to understand." The family's reading materials include *Discover Magazine*, *Popular Science*, and *National Geographic*, which Gitanjali says feed her project ideas. "My mom, on the other hand, introduces me to daily news that impacts people and has always played games with me and my brother about day-to-day problems and finding creative solutions."

Following her winning the 3M contest, she appeared at the 2018 MAKERS conference in Los Angeles. Gitanjali explained her device to an audience largely composed of entrepreneurial and professional women. Acknowledging a couple of her role models, she mentions Shafer of 3M and Hernandez-Ruiz at Denver Water for helping her learn about innovation and for showing what women can do in STEM.

After a standing ovation, the hosts called her back onstage. They challenged the audience to give her an additional $25,000. This would match her 3M contest winnings and kick-start her efforts to turn Tethys into a full product. Within a couple of minutes, they reached that goal.

Gitanjali has granted many interviews and appeared on many national and local news broadcasts. She's an eloquent role model for ingenuity, persistence, and other young women considering STEM fields. "I definitely want to be part of that movement and getting more females involved in STEM." She's proud to help break down the old stereotypes about the role of women in science and especially women of color. "You have to develop resistance to [those stereotypes], and perseverance comes out of experience."

Gitanjali has created workshops to share her innovation process with other teens. These hands-on labs plunge participants into the invention process she's developed: observe, brainstorm, research, build, and communicate. She offers her workshops through schools, museums, STEM organizations for girls, and international sponsors including the Royal Academy of Engineering in London and the Shanghai International Youth Science and Technology Expo group. Gitanjali discovered that many young people just don't know how to begin innovating. She's proud to spark creative ideas in others. "That means one more person in this world wants to come up with ideas to solve problems."

I think that being an inventor-slash-scientist is pretty much like [being] a superhero.

—Gitanjali Rao

When asked about her long-term plans, she always mentions her intentions to study at MIT. "I think that being an inventor-slash-scientist is pretty much like [being] a superhero," she says. "What do superheroes do? They save the world. They save lives. And what do scientists do? They could potentially save the world, they could potentially save lives. So, we are pretty much superheroes."

WILLIAM KAMKWAMBA AND THE WINDMILL

"Before I discovered the wonders of science, I was just a simple farmer in a country of poor farmers," says William Kamkwamba in his 2009 TED Talk.

William was born in the tiny farming village of Wimbe in Malawi. This small African country is one of the poorest nations in the world. The average annual income in Malawi is about $13,000 USD and the life expectancy is about fifty years. Over 80 percent of the population lives on what they can grow on their small farms. Malawi suffers from both droughts and floods. Either not enough rain falls to water the crops or else so much that the crops drown. "Farmers here have always been poor, and not many can afford an education . . . from the time we're born, we're given a life with very few options," William says.

William lived with his father, mother, and six sisters. His family farmed a small plot of land at the edge of the village where they grew corn and tobacco. It was hard, backbreaking work with the hand tools they used to till the earth. Everyone worked from early morning to sundown during the growing season. "My sisters and I would wake up

Most people in Malawi work in agriculture, growing tobacco, sugar, tea, legumes, and soy. Malawi's economy is also supported by its manufacturing, tourism, and STEM industries.

before dawn to hoe the weeds, dig our careful rows, then push the seeds gently into the soft soil," William recalls. "When you live on a farm, you need all the help you can get."

When the family was not laboring in their fields, they gathered firewood, carried water, made clothing, cleaned their homes, prepared food, and cooked meals over small wood-burning stoves, all by hand. After dinner and after the sun went down, William's family, like most all the villagers, went to bed around 7 p.m. They had no electricity, no running water, and no indoor toilets.

Malawian farmers, like William's family, lived from harvest to harvest. Any problem with bad weather, the fertilizer mix, or low-yielding seeds can push these families from barely getting by into hunger. A poor harvest meant that the typically large families in Malawi must exist on their small supplies of stored food. William's family often had to skimp on meals during the periodic famines.

William's family valued education, and they spent much of what little money they had to send him to the village school. Malawi has no free public education past primary school, so few Malawians attend middle school or high school. Even when the harvest was bountiful, continuing education was an enormous expense for the family. Tuition took much of their remaining cash after they bought the necessities that they could not grow or do without, such as kerosene or soap.

Radio Repair

William discovered his love for science around the time he turned thirteen years old. "From the first time I heard the sounds coming from the radio, I wanted to know what was going on inside."

Along with his good friend Geoffrey, "I started taking apart some old broken radios to see what was inside," William recalls. "I'd stare at the exposed circuit boards and wonder what all those wires did, why they were different colors, and where they all went . . . who'd arranged them this way, and how did this person learn such wonderful knowledge?" William says, "We began figuring out how they worked so we could fix them."

Through trial and error, with no manuals or teachers, the two friends learned a great deal about how radios worked. They would disconnect parts and see what happened. They also wired the parts in different ways. Then they discovered that the biggest part inside the radio produced static sounds when connected to power. This part, the circuit board, contained all the wires and bits of plastic. They wondered about some small parts that looked like beans. By disconnecting one, they found the volume of the radio dropped significantly and they reasoned that the beanlike transistors regulated the power to the speakers. When they experimented with the antennae, they saw that the AM bands used an internal antenna, while the FM bands needed an external antenna that was not obstructed by trees or buildings. Later, William would read about the longer AM radio waves and the shorter FM waves, closing the loop on this hands-on discovery.

They experimented with ideas that came to them as they went along. Using this trial-and-error method, William and Geoffrey sacrificed many radios in their pursuit of knowledge. With no proper tools, they improvised a soldering iron by heating a thick piece of wire over the kitchen fire. With little spare money to buy batteries, they scavenged discarded batteries from the trash at the local trading center. They connected several together to extract the remaining power. They quickly discovered the differences between the positive and the negative poles and figured out how to hook them up, even without knowing the names or the theory yet.

"After we learned from our mistakes, people began bringing us their broken radios and asking us to fix them," William says. "Soon we had our own little business."

William also grew fascinated by all kinds of mechanical devices and machines. He asked everyone questions and would examine any mechanism he encountered to see how it worked. One evening a neighbor rode up on his bicycle to see William's father. As he approached, the bike's headlamp shone brightly but turned off as soon as the man hopped off the bike. Intrigued, William asked the neighbor why the lamp went off. He listened intently as the man explained that the turning of the wheels turned the dynamo and powered the lamp. When he stopped pedaling, the lamp went off.

William had seen bikes with dynamos before but had given little thought to these devices that looked like small metal bottles attached to the wheels. When the neighbor went inside, William hopped on the bike and pedaled until the light came on again. Stopping, he flipped the bike over and examined the lamp and the dynamo. He found wires leading from the lamp back to the dynamo and saw that the dynamo had a small metal wheel that pressed into the tire. As the bike tire moved, the metal wheel of the dynamo spun, and then the light came on.

"I couldn't get this out of my head. How did spinning a wheel create light? Soon I was stopping everyone with a dynamo and asking them how it worked," William says. The idea of producing electricity from a spinning bike wheel fascinated him. One afternoon he borrowed his neighbor's bike again and flipped it over. Then he connected the wires from the dynamo to a radio with the batteries removed. He and Geoffrey cranked the pedals but nothing happened.

William experimented by reattaching the wire to the headlamp and turning the crank. The light flickered on. Then he took the batteries from the radio and connected them to the headlamp by a separate wire. The light came on again. William was puzzled. Both the bulb and the dynamo worked, and the batteries from the radio worked to light the bulb. Why wouldn't the dynamo power the radio? As the two baffled friends looked for an answer, Geoffrey had an idea. Pointing to a socket on the radio labeled AC, Geoffrey suggested connecting the wires from the dynamo to that socket.

Unhooking the batteries, William pushed the headlamp wires into the AC socket. Music began pouring from the radio. The two friends shouted in excitement and swapped turns dancing to the music and cranking the pedals to keep the radio going.

"Without realizing it, I'd just discovered the difference between alternating and direct current. Of course, I wouldn't know what this meant until much later," says William. Before long, they tired of pedaling the upside-down bike. "So, I began thinking, *What can do the pedaling for us so Geoffrey and I can dance?*" William recalls.

This successful experiment with generating electricity inspired William. He wanted to make more electricity. William's family were among the 98 percent of Malawians who had no electricity. With no lights, William could not study at night or work on his radio repairs. Just like everyone else, William went to bed shortly after sunset.

Hardships

In the best of times, the life of a farmer in Malawi was full of hard work and uncertainty. Even after a good harvest, most families had little money left after buying seeds and fertilizer for the next year. Many years, the corn supplies that families had stored up ran low in the times before the harvest. People scrimped and cut back their meals until the harvest came, which could be a month or more away.

After a poor harvest, families tightened their belts further, cut back even more on meals, and endured until the harvest. "We call this period 'the hungry season,'" William says.

But December 2000 was a terrible year. The cost of fertilizer was much higher when the government deregulated the price. William's family could not afford to buy any fertilizer that year. Then the rains came much harder than usual and flooded the fields, washing away houses and livestock along with the seedling corn. As suddenly as the rains came, they stopped, and a long drought began in Malawi. William's family only harvested five bags of corn that year. They sat in one small corner of the storeroom. In a good year, the storeroom was full to the ceiling.

After this disastrous harvest, food ran short. Many people of Malawi went hungry, and William's family began skipping breakfast. As the famine became very severe, starving people roamed the countryside trying to find day work or pleading for food. By then William's family was only eating dinner, their one meal of the day. William remembers the anguish of his parents who had so little food to feed their family of nine. Everyone always went to bed hungry. William's mother told her children to drink lots of water with their cornmeal porridge so their stomachs would feel fuller.

Near the end of the famine, a cholera plague swept the countryside, and many people, already weakened by hunger, died. These were truly desperate times for William, his family, and the other people of Malawi.

During this devastation, William and his friends graduated from the small, local primary school. William looked forward to middle school where he could study science. "I'd decided I wanted to become a scientist—not just *any* scientist, but a great one."

But William's family could make only the down payment on his tuition. One day not long after classes started, the head teacher told William and some other boys that they must go home. They could not come back until they brought the cash to pay their tuition.

But there was no tuition money for William during this famine. Rising at 4 a.m., he worked all day in the hot sun side by side with his father. He knew it would be a long time before his family had the money for school again.

The Library

When he was not in the fields, William went to the trading center to play games and socialize with other boys and men. "But chess and bawo [mancala] weren't enough to keep my mind occupied. I needed something to trick my brain into being happy. I missed school so terribly."

One day William had an idea. If he couldn't go to school, he would make his own school. The Wimbe Primary School had a library with books donated from the United States. The librarian knew William and welcomed him to the library. As she showed him the collection, the number of wonderful books to study awed him.

He sat on the floor and read most of that day. When he left, he had a stack of borrowed books, including the same textbooks his friends were studying in school.

Following his "alternative school" plan, he would go to the library when he was not working in the fields. He read for most of the morning and then read more at home in the afternoon. He also borrowed school notes from his good friend Gilbert, who was still in school. When he didn't understand something in the notes, Gilbert would explain it to him. At night William studied until it was too dark to see. But he kept at it, working the fields or reading in the library by day and studying by night. His family helped and supported him all they could.

Studying Science

One Saturday at the library with Gilbert, he found a book called *Malawi Junior Integrated Science*, a textbook used in school. As they read, William came upon a picture of the huge Nkula Falls with a hydroelectric plant run by the Electricity Supply Corporation of Malawi (ESCOM), the state-owned power utility.

He knew that the water from the Shire River, the biggest in Malawi, flowed until it reached Nkula Falls and the ESCOM plant, which produced electricity. But he had no idea how the plant made electricity. As he read, William learned the falling water turned a turbine that made the electricity. This seemed to William like the same concept as the bicycle wheel turning the dynamo. Could he somehow use a dynamo to make more electricity?

Another day, he came upon a textbook for older students called *Explaining Physics*, written in English, a language William was just learning. This book was much more difficult for William to read. Long, complicated words and phrases, as well as technical terminology, filled its pages. Determined, he struggled with the text over the next week and managed to figure out enough words to grasp the gist of the content.

William developed a strategy to decipher the technical terminology. If he wanted to understand an illustration, he'd look at the label for the figure, say figure 10, for example. Then he'd scan the text until he found where it discussed figure 10, and then he would study the surrounding sentences. Going back and forth, he'd used the text to understand the illustration and then the illustration to understand more of the text. This was a slow process, but his hands-on knowledge of electronics helped.

William connected his experience with repairing radios and dissecting machinery to the scientific knowledge in the book. What he did informed what he read. For example, magnets fascinated him. He knew about magnets from taking apart old electric motors. As he read about the scientific properties of magnets, he linked that with his own hands-on discoveries. From playing with these magnets, he knew that magnets both pushed and pulled. The book illustrated the invisible lines of force and the magnetic field surrounding a magnet. The lines of force looked like the wings of a butterfly to William. He read about making electromagnets by winding wire around an iron core and attaching the ends of the wire to a battery.

He also learned how electric motors work. William explains, "A coil of wire on a shaft is housed inside a large magnet. When the coil is connected to a battery and becomes magnetized . . . the push and pull between the two magnetic fields causes the shaft to spin." This spinning motion of the shaft can power appliances "like the fan we use in hot weather."

Then he found a new idea in the book. He read that if a device with a power source can create a spinning motion, as in a fan, then running the process in reverse would generate electricity. In the same way that falling water at Nkula Falls spun the blades of the turbine, spinning the blades of a fan would cause the coiled wire inside the device to rotate. This rotary motion would cut through the force field of the magnet inside the device and produce electricity. Scientists call this effect electromagnetic induction.

This same basic device, William learned, can both use electricity to produce motion, which makes it a motor, or use motion to produce electricity, which makes it a generator.

The best example of a generator in the book was a bicycle dynamo creating current for the headlamp as the rider cranked the pedals. "*Of course*, I thought. This is how spinning motion generates power, both in dynamos and in the hydro plant!

"I can't tell you how excited I thought this was. Even if the words sometimes confused me, the concepts illustrated in the drawings were clear and real in my mind . . . [they] made perfect sense and needed no explanation," William marvels. All at once, he could "see" the principles of magnetism, how magnets induced electric current, and how AC differed from DC. His work with radios and machinery had created a jigsaw puzzle map of these principles in his brain, he says. Upon discovering these new ideas in the science books, the missing pieces of the puzzle snapped into place. Elated, William kept the book for a month and studied it each day.

When he got stuck, he asked the librarian to look up words for him. *What does "kinetic energy" mean? And "resistor"?* Mrs. Sikelo, the helpful librarian, told him that he had gone further than his classmates in school with the science curriculum. William already knew that he was ahead and that he really needed to learn this material.

The Windmill Vision

One day while looking for a dictionary, he noticed a book stuffed way back on the shelf. "*What's this?* I thought. Pulling it out, I saw it was an

American textbook called *Using Energy,* and this book has since changed my life." The book's cover showed a row of graceful windmills, although at that time he had no idea what a windmill was. To him they looked like the toy pinwheels he and Geoffrey made as children.

He opened the book, flipped to the chapter titled "How Do You Generate Electricity?" and began to read. "Energy is all around you every day," the book said. "Sometimes energy needs to be converted to another form before it is useful to us. How can we convert forms of energy? Read on and you'll see." He read on.

William learned that windmills provided the energy to pump water and grind grain. The book said that many windmills grouped together, like those on the cover, could provide as much electricity as a power plant.

Suddenly, William saw that the wind drove the blades of the windmill the same way it spun the toy pinwheels. He remembered what he learned from his neighbor's bike and the book *Explaining Physics.* The dynamo powered the headlamp through the pedaling of the rider. "*Of course,* I thought, *and the rider is the wind!*" He vividly pictured the process of the wind spinning the blades of the windmill, which rotated the magnets inside the dynamo. The rotary motion of the magnets produced electricity.

Fired up by these thoughts, William realized that with a windmill his family could have electric lights. They would no longer have burning eyes or gasp for breath because of the fumes from their kerosene lamps. He could stay awake reading and repairing radios into the night.

But even more important than lighting his room, he realized a generator could pump water from their well to the crops. Even if the rains did not come, the crops would grow, and his family could eat. There would always be enough maize for three meals a day. *Wow,* he thought, *we could even grow two maize crops a year!* "With a windmill, we'd finally release ourselves from the troubles of darkness and hunger," he says. "A windmill meant more than power, it was freedom.

"I decided I would build my own windmill."

The First Windmill

In his mind, William saw the windmill he planned to build. He knew that building a small model first would give him the opportunity to experiment and figure out the mechanism before attempting a full-scale windmill. From examining the pictures in the book, he knew he'd need windmill blades, an axle the blades could spin on, and a generator, along with some wires. He gathered these materials from around the house and also from parts scavenged at the village dump. He even swiped an old plastic lotion jar his sisters were using to play cricket.

By sawing off the bottom and spreading the sides of the jar, he made the blades. Nailing the blades to a pole in his yard, he noticed that the wind barely turned them. Realizing they needed to be longer, he made blade extensions out of old polyvinyl chloride (PVC) pipe. He softened the pipe over his mother's kitchen stove. Then, using tools he made from old bicycle spokes, he cut and drilled the softened pipe, molding the pieces into the shapes he needed. He finished before his mother caught him in her kitchen and banished him to the fields to help his father.

As he worked in the fields, he thought about his windmill project. *How can I get a generator?* Then he realized from his reading that the generator for the model didn't need to be very big for his test.

With this thought, he knew just where he could get one. Remembering an old cassette player that he and Geoffrey had worked on, he took out the small electric motor. He hooked this up to his windmill so it would act like a generator and produce electricity. Rather than use electricity from a battery to operate the cassette player, the mechanism would make electricity using the power of the wind.

With his homemade tools, William assembled his small windmill. He attached his PVC "pinwheel" to the lotion jar base. Needing a rubber washer, he shaped one from the rubber heel of an old shoe, rather than buying one at the trading center. William's inventing required him to improvise parts and tools frequently.

Changing Energy

Electric motors and electrical generators are essentially the same electromagnetic device. Both consist of a rotor wound with wire inside a magnetic field created by an electromagnet. This device turns one form of energy into a different form.

Motors turn electrical energy into mechanical energy. A simple house fan uses electrical energy from the wall socket to power a motor that turns the blades of the fan. Generators turn mechanical energy into electrical energy. William used the energy from moving air to produce electricity. His windmill fan blades turned the rotor inside the bicycle dynamo.

William also used this idea when he was testing his first windmill. Taking an old cassette player, he removed the battery-powered electrical motor that spun the cassette tape (electrical energy to mechanical energy). Then he hooked the device to his windmill so that the wind would spin the rotor to produce electricity to play music (mechanical energy to electrical energy). The same device can work in both directions.

Scientists think of magnetism and electricity as two parts of the same basic force of nature: electromagnetism. Charged particles (electrons) moving inside a wire produce a magnetic field around the wire. This is why electromagnets work. Conversely, a moving magnetic field can push electrons along a wire to produce an electrical current. (See the sidebar on pages 29–30.) William learned all of this by combining what he discovered through experimentation with what he read in the Wimbe Primary School library.

Then he connected the cassette player to the windmill. If it worked, the spinning windmill blades would spin the magnets in the motor, turning it into a generator. The generator would power the cassette player with wind power instead of battery power.

Ready for a test, William and Geoffrey brought the windmill and generator out into the cornfields. William hooked up their generator to Geoffrey's radio and then held the windmill up to the wind. The wind blew. The blades turned. And then the music played from the cassette.

"'You hear that, man?' I screamed. 'We did it! It actually worked!'"

Going Bigger

Then William knew he could build the windmill he envisioned. "The model had already revealed itself in my mind," he recalls. He also knew what parts he needed for this new windmill. William visualized a bike frame as the backbone. He would fasten the blades to a sprocket and chain. As the wind blew, the back wheel would turn and set the dynamo rotating. After the success of his model, he was confident his plan would work, though he had none of the materials he needed or the money to buy them. Time for more scavenging and improvising.

Over the next month, William spent many hours in the junkyard near the middle school he attended so briefly. With a purpose and a plan, he found the junkyard a bountiful source of materials and ideas. He explored and experimented with the discarded treasures he found there.

From an old tractor, he found a big radiator fan for the blades and a rod that would make a great axle. Finding ball bearings of the right size and some grease took longer. Time and determination were his currencies.

As he worked, he surveyed all the broken and scattered machinery and tried to visualize how they worked. He missed his friends, but most of all he missed learning. One day as he left the junkyard, he looked up at the school and called out. "'Look out,' I said. 'Your man Kamkwamba will be back soon.'"

The next year, the rains came, and the harvest was one of the best anyone could remember. As the new crop came in, William and his family could go back to eating three meals a day. As the family recovered from the famine and life returned to more normal times, William thought about school again. His family, however, could still not pay the tuition after they paid off their creditors, who lent them money for food during the famine.

So instead, William sneaked *into* school. He'd wait for the morning assembly and then come in the back. He kept his head down and said nothing, did nothing to draw attention to himself. Finally, however, the head teacher caught him and sent him home.

William worried. Would he be a subsistence farmer forever, just like his father, and his father before him, "another poor Malawian farmer laboring in the soil"? Would his life be one of alternating famine and hunger and then seasons of good harvest? "Thinking about it now scared me so much I wanted to be sick."

William Has a Project

William continued to work in the fields with his father. When he could get a break, he went back to the junkyard and continued collecting the parts and pieces he would need. Though he finally had most of his parts, William still needed a frame for the device. Then he realized that he had just what he needed at home.

William's father had a broken bicycle with only one wheel. Since he hoped to fix the bike someday, he didn't want William to take it apart. William then shared his dream with his father and asked if he could try to make that dream come true. He explained what it could mean for their family: lights for the house, an electric pump to water the crops, and an extra harvest each year. But best of all, they would never go hungry again. After considering this a bit, his father relented and agreed to support William's project.

William put the broken bike in his room with all of his scavenged treasures, which was beginning to look like a junkyard itself. But he kept his hard-won treasures since, like tinkerers everywhere, he never knew what he might need next.

His father let him stay home from the fields to study and build his windmill. When his sisters complained to their father about William not working, he told them that William was working on a project.

As William continued to mine the junkyard for parts and pieces for the windmill, the boys at the school jeered at him. They teased him mercilessly about digging in the garbage. William was a madman, he was smoking marijuana, and he was wasting his time, they taunted.

People from the village also called him misala (crazy) and gossiped about the lazy boy who played with toys all day long. Throughout the ridicule, his father continued to support him. "Leave the boy alone . . . let's see what he has up his sleeve."

Getting a Dynamo

William continued making his windmill. He and Gilbert secretly dug up old PVC pipe from around Gilbert's home to make bigger, extended blades for the radiator fan to catch the wind better. He again used his handmade tools and his mother's kitchen stove to melt and mold plastic.

But he still needed a generator. He tried to make one. He needed a long continuous strand of wire but could not find one long enough. The bits and pieces of wire he connected didn't work. Nor could he find the other parts he needed, and there was still no money to buy one.

One day while walking around the village with Gilbert, William noticed a stranger with a dynamo on his bike. Gilbert asked the man how much he would sell the dynamo for. *Two hundred kwachas,* the man said. Gilbert pulled two red hundred-kwacha bills from his pocket and gave it to the man. *Let's build this windmill!* Gilbert said.

"It was like adding the last piece of the great puzzle of my life," William recalls.

Electric Wind

The next day, starting early, he worked until he completed assembling his windmill. Looking up, he was surprised to see the sun going down.

After dinner, he heated water in the kitchen for his bath, washed up, and then quickly fell asleep.

William awoke the next morning full of anticipation. As the sun rose, he put a bamboo pole in the ground as a stand for his windmill. Geoffrey came along at just the right time to help hoist the windmill and fix it to the pole.

When Geoffrey pulled the pin that held the blades stationary, the blades began to rotate. They spun slowly at first, then faster and faster. With a gust of wind, the blades spun so fast that the bike chain snapped.

With the blade mechanism working, did the dynamo make enough current to power lights? After fixing the chain, William got his father's radio and connected it to the dynamo.

Again, the wind blew, the blades turned, the wheel turned . . . and then music came from the radio. But only for a few seconds. Then smoke began pouring from the radio and it nearly caught fire. They quickly disconnected the wire and the smoke stopped. With the fire danger averted, Geoffrey worried about the consequences of the ruined radio. William, however, was too excited to care and marveled at how much power the windmill made to produce the smoking remains before them. A big windmill could clearly generate the electrical power William wanted.

But why did the radio burn out? What went wrong?

Pondering the problem, William again searched for an answer from his science books. Then he realized that his father's radio could only handle six volts of power and the dynamo was producing about twelve volts. The radio's circuits overloaded and burned out.

William knew he needed something to reduce the voltage. He remembered reading that electrical voltage dissipated as electricity traveled long distances over wires; that was the reason transformers were needed to increase the voltage. *What if,* he pondered, *I could make the electricity travel further along a wire to decrease the voltage?*

He went to his room and found an old motor with the wire he needed. He wound the wire around a core. Adding this device to his windmill, he reduced the voltage enough to avoid burnouts.

A Tower for the Power

With his windmill assembled and working, William wanted to raise the windmill above the surrounding trees so it could produce more power. Gilbert and Geoffrey agreed to help him build a tower.

They chopped several trees from some nearby woods and nailed them together in a pyramid behind William's bedroom. When they finished at sundown, the tower stood 16 feet (5 m) high. William remembers thinking that although it was tall enough and strong enough, it swayed a bit like a drunken giraffe. No matter. Tomorrow they would raise the windmill. With a day of strenuous work behind them and another before them, they headed to bed. William dreamed of windmill-generated electricity and the hope it would bring to his community.

Around seven the next morning, they began hoisting the 90-pound (41 kg) windmill up the tower using the thick wire his mother used for a clothesline. Sweating and groaning, the three friends secured the windmill in place at last. Then they happily surveyed the windmill, which was ready for a test run.

When William was atop the tower, he noticed something in the distance. After seeing the tower go up, people were coming to see what was going on. About a dozen villagers came to see this curiosity.

One man shouted up to William, *What is it?* Since there was no word in their Chichewa language for "windmill," William shouted back, *It's electrical wind! It generates electricity from the wind.*

The gathered crowd buzzed. *This is impossible! This is the crazy boy from the junkyard. His mother must be so ashamed. It will never work.* William saw a larger, growing crowd of about sixty people and his family watching from below. He attached the small light bulb from the bike headlamp to a cord and socket, and then he released the pin on the windmill.

Pointing at him and mocking, the crowd settled down to watch William fail. *Here goes,* William thought as he released the blades. "The chain snapped tight against the sprocket, and the tire spun slowly, creaking and groaning at first. Everything was happening in slow motion.

One of William's first windmills provides electricity to the buildings nearby.

I needed it to go faster, immediately. 'Come on,' I begged. 'Don't embarrass me now.'"

The blades begin to turn, just barely. Then a bit more. Finally, the blades whirred. William held the small bulb aloft for all to see. A gust of wind almost blew him off the tower, but he hung on.

"I held the bulb before me, waiting for my miracle. It flickered once. Just a flash at first, then a surge of bright, magnificent light. My heart nearly burst."

Someone in the crowd shouted for everyone to look. Everyone turned their heads to see what had happened. Though it seemed miraculous, William had made light.

"It was a glorious light and it was absolutely mine! I threw my hands in the air and screamed with joy. I began to laugh so hard I became dizzy." William now dangled precariously from the tower, brightly burning bulb in hand like a modern Prometheus. Surveying the crowd, he saw the wide incredulous eyes and the gaping mouths.

"'Electric wind!' I shouted. 'I told you I wasn't mad!'"

With that the crowd began one by one to clap and cheer. *Outstanding work, William. You did it. We didn't think it could be done, but you did it.* People crowded around Gilbert and Geoffrey, wanting to hear more details. Neither of William's friends could contain their glee. William stayed in his perch for about half an hour, letting every incredible moment soak in. Finally, the bulb got too hot to handle, and only then did William climb down.

Other Tech Projects and Inventing

William basked in his success. The next day he strung small car headlamp bulbs around his room, lighting it up as he had dreamed. At last he had his own private lighted space.

In the days that followed, William wired the rest of his family's house and installed a car battery to store power for when the wind didn't blow. He built a transformer to charge his cousin Ruth's cell phone, applying what he'd learned from the chapter on transformers in *Explaining Physics*. After some patched wiring in his wood and grass roof sparked and nearly caused a fire, he devised a circuit breaker to prevent fires in the future. Using the same principle as a doorbell, he balanced a small bar magnet between two electromagnets made from nails and copper wire. A surge in current, as happens with a short circuit, would upset the balance, pull the bar magnet to one of the electromagnets, disconnecting a small wire, and breaking the circuit. He and Geoffrey also made a radio transmitter that could broadcast about 300 yards (274 m). On this inventing and engineering spree, William figured out how to build these devices from his physics books and his experience taking radios apart.

One evening, William walked into his living room and saw his mother crocheting in the light while his father and sisters listened to the radio. He asked his father if he liked not having to pay ESCOM for the light. His father said, *Yes, but I'm even happier because my son made this possible.*

Even though his inventing was going well, William missed going to school and wondered what was next for him. He wanted to do something and to make a difference. "If I can teach my neighbors how to build windmills, I thought, what else can we build together?"

Getting Noticed

People came from miles around to see the windmill. It even attracted businesspeople visiting the area. Hartford Mchazime, a government education official, came to visit William and his windmill.

Impressed with William's work, Mchazime asked, *But who taught you?*

I read books and taught myself, William told him.

Mchazime immediately brought media attention to William's accomplishment and set about sending him to school.

> If I can teach my neighbors how to build windmills, I thought, what else can we build together?
>
> *—William Kamkwamba*

After that, things moved quickly. William was soon going to a new school, and he received an invitation to speak at a TED conference in Tanzania. TED conferences gather leading thinkers, influencers, and funders to discuss global issues and innovative ideas. TED opened up a whole new world for William, full of scientists, inventors, engineers, and venture capitalists. Though English was difficult for him, he stood onstage and talked about being forced to drop out of school. He told the audience about going to the library to study and finding the science books about windmills. Then he shared the story of his goal to build a windmill of his own. "I try and I made it!" he said. After the talk, William felt a kinship with the people at TED. These people had ideas and resources to help people.

With the money and support he raised at TED, William went back to Wimbe and strengthened his windmill, rewired his home, and installed solar panels to supplement the windmill power.

Sponsored by his new friends from TED who wanted to bring his story to a wider audience, William toured America. He appeared on talk shows, visited New York City, and attended the CES, a consumer electronics convention in Las Vegas. He also visited the actual windmills in California pictured on the cover of *Using Energy* and, in Washington, DC, he met the textbook's author, Mary Atwater. "Like me, she didn't listen to what others said. And, also like me, Dr. Atwater had a father, Mr. John C. Monroe, who believed in her dreams and saved money to send her and her siblings to college." Atwater was a Black woman who loved science, and with the support of her father, she persevered against racism and sexism to earn her doctorate in science education. Atwater then went on to teach at the University of Georgia. There she wrote the book that ended up in Wimbe, Malawi, and forever changed the life of a boy who desperately wanted to learn science.

Giving Back

On one talk show, William mentioned needing to study for his college entrance exams. Within days, he had many offers from prestigious colleges. William chose Dartmouth College because of their creative engineering program and the welcoming people he talked to. There, he met his future wife, Olivia, in a math class. Following college, William accepted a fellowship from IDEO, an innovative design and engineering firm. IDEO runs a nonprofit organization that collaborates with developing nations and communities around the world to solve local problems with simple, cost-effective technology. At IDEO, William worked on projects such as a low-cost sensor that lets rural farmers know when their crops need more water or fertilizer.

William also continues to help out his family, his friends, and his hometown. Back in Wimbe, he had new metal roofs put on all the houses and added solar cells for power. Fulfilling his first vision, he made a water pump for their well so his mother wouldn't have to carry water. He deepened her well and installed a tap on the well. Then his mother and anyone from the village could get water without having to carry it so far. The pump also watered the cornfields, banishing the fear of drought.

William's parents still live in Wimbe, Malawi. A combination of windmills and solar panels provides electricity to their home. Moving Windmills, among its other initiatives, helps install solar panels on Malawian homes.

William wholeheartedly gives back to those who helped him and to others who need his help. William, along with his friend Tom from TED, founded the organization Moving Windmills, which directs funds for improving villages and schools near his home. For example, in partnership with buildOn, which helps build schools in developing countries around the world, William led the expansion of the Wimbe Primary School. The people in the village of Wimbe made all the bricks and built the new schoolrooms. He also aided Kachokolo Secondary School, the site of much grief for him, by giving it new computers, desks, and an internet connection.

Through his nonprofit organization, William is starting the Moving Windmills Innovation Center to provide a hub for young African inventors and engineers to use design thinking to solve problems in their community. Located in Kasungu, Malawi, near his hometown, the center will provide inspiration, mentorship, and tools. Young innovators will come here to develop design and building skills and take away a sense of personal agency to tackle problems. In creating the center, William tells of being inspired by his grandmother who made her own bricks and built her own house.

"I see there are so many talented young people all over the world," William says. "But sometimes they don't have a chance. If they can have space to work with me, they can solve real community problems." William tells those who would emulate him, "Don't doubt yourself, and when you are trying to do something, don't be afraid of failing."

William also helped his family and friends by founding several companies. Since his father could grow more corn, William bought him a couple of pickup trucks and built him a corn-grinding plant to run. He created a solar panel company for Geoffrey. He helped Gilbert found a video store where Gilbert also produces and records African singers.

William is dedicated to making life better for Malawi and all of Africa. "Sure," he says, "it would be nice to stay in the States, find a good job in Silicon Valley or New York, and make a lot of money. But that's not me . . . my heart and life's work belong in Africa."

He and Olivia, who is completing her doctorate in science education, plan to live and work in Malawi, where she can teach and do her research on education. They may also spend some time in the US with William's many friends.

William's parting advice to fellow inventors and innovators: "Trust yourself and believe! Whatever happens, don't give up!"

4

AUSTIN VESELIZA AND THE TALKING GLOVE

On a sunny day at a Trader Joe's grocery store in San Mateo, California, one of the checkout lines stalled. Austin Veseliza, a tall, quiet ninth grader, looked up to see the problem. The customer at the head of the line was struggling to communicate with the checkout clerk. The customer seemed to have a condition that affected her speech. Her rapid hand motions meant nothing to the clerk who only shrugged and pointed. Austin could not understand the customer either, nor could others nearby. The customer's gestures grew wilder and more emphatic, her face reddening. Although the clerk was watching her closely, he still shrugged in confusion. Other customers waiting in line behind her became restless. Finally, she tried to write a note on paper, but frustrated and deeply embarrassed, she gave up and fled the store.

Austin, troubled by what he saw, couldn't stop thinking about the Trader Joe's customer. What would she do now? Did this happen often? For Austin, communicating with others was easy and quick. For this woman, communication could obviously be very difficult. She was

Author's Note

Writing respectfully, yet clearly and concisely, on the topic of speech conditions is difficult, as there is little consensus on terminology. After considerable research, consultation, and thought, I have adopted the person-first convention and aimed for non-stigmatizing language for this book. The main text refers to "speech difficulties" or "speech condition," while the sidebar, which deals with medical aspects of speech, uses the convention of "speech disorder" in discussing physiological causes.

trying, but because no one could understand her, they didn't seem to know what to do. *There must be a better way,* Austin thought. Could he invent something to help this woman and people like her have an easier time communicating with strangers? "I felt like I had stumbled onto a chance to make a difference," he said.

Early Inventing

Austin was an inventor. As far back as he could remember, the stories of Thomas Edison, Leonardo da Vinci, and other famous inventors inspired him. He wanted to be like these inventors who developed astounding technologies. In his free time, he tinkered and built contraptions from scrap wood and cardboard. What he saw in his mind took physical form. His parents' garage held many of his earlier creations in various stages of completion.

When he was eleven years old, he and two friends from school entered a science and technology contest called eCYBERMISSION. "My friend, Ben, said, 'Hey, I want to do the science fair team. Do you want to join?' and I said, 'Yeah.' So, it was me, Ben, and Johnny," Austin recalled. The three friends set to work, though, looking back, "we just didn't know what it was going to be at all. We [just] knew we had deadlines." All the parents supported them, but they let the team own the project. "Ben's mom, especially, was helping us through the process," Austin said. She helped them plan their work and calendar their deadlines. "She would make sure we were meeting and actually doing the work and not just goofing off. But then she was hands off after that in terms of doing the actual work."

They chose improving sports helmets to reduce head injuries as their project for the eCYBERMISSION competition. The three friends researched and tested many padding materials until they found the right combination. "There was also some conflict at times because we spent so much time together. We were at each other's houses maybe two days a week after school and a full day on weekends for months and months," said Austin. "There was disagreement on how we should approach each given problem a lot of the time. Which material should we test? How should we get those materials? How should we test them? It was three different people with different ideas about how we should do every step."

To their astonishment, they won first prize for the sixth-grade division that year in their local eCYBERMISSION contest. "We had no idea at the time it was going to turn into going to the national contest," Austin recalled. They won first prize at the national contest as well, winning a trip to the White House to meet President Obama and shake his hand. "I can't believe it happened. It almost feels like it happened in a different world, if that makes sense." In addition, they met celebrity scientist Bill Nye, Adam Savage and Jamie Hyneman from *MythBusters*, and Dean Kamen, who founded the FIRST Robotics Competition. The pictures taken that day amid the swirl of people and activity show the young inventors' simple joy. This contest experience only boosted Austin's zeal for invention and his desire to improve the tools of daily life. "It made me feel capable, and I was on a good path for me," he said. "I was playing to my strengths, and I was doing good work. It definitely spurred me to keep going, for sure."

School Design Project

At just the time of his experience with the Trader Joe's shopper, Austin needed a project for school. He was taking an engineering class in school called Design Thinking. "I was looking for some design need out in the world. My eyes were open for that," he said. "This felt like a very human problem. It's something that [people with speech difficulties] would be forced to deal with every day. These types of needs are very motivating to me. . . . It felt like it mattered."

But would this make a good class project? "I've heard that the strategy for being able to do your best work is to find the intersection of what you value in the world and what you're excited to work on," Austin said. And he added with a laugh, "And with what you can actually do. . . . Those factors led me to choose this project."

I've heard that the strategy for being able to do your best work is to find the intersection of what you value in the world and what you're excited to work on.

—Austin Veseliza

Austin wanted to find a way for the Trader Joe's grocery shopper to communicate more successfully in situations like that of the checkout line. What would work? Everyday life has many unplanned interactions. People who have speech difficulties would face more barriers in such unexpected situations. Effective communication is vital to living independently. In an emergency, it might even be lifesaving.

Many people with a condition that impairs speech use sign language, a language that relies on hand gestures rather than voice to communicate. This allows them to converse with others who can sign. But most people never learn sign language, and it isn't taught in most public schools. The widespread lack of understanding of sign language leads to frustrating communication barriers for everyone, and people with speech difficulties often have to rely on other methods to be understood. Recognizing the problem, Austin used design thinking to begin looking for another way.

He did a quick assessment of the options already available to people with speech difficulties. "I started by looking at what existed out there already," he said. His preliminary research showed weaknesses in the existing solutions.

Problem-Solving, with a Twist

Design thinking is a way of solving problems. It centers the people who experience the problem and focuses on what new ideas might work as solutions. Design thinkers are flexible, creative, and curious. Many schools offer courses in design thinking, mixing problem-solving with people skills.

People often connect design thinking with STEM labs, but design thinking works in any field. Corporations, universities, new tech companies, and social action groups apply design thinking when they need an innovative solution. This method works for everything from making new products, such as Apple's Mac computer or Tesla's electric cars, to better ways to get food to hungry people. Many of the inventors in this book use design thinking or follow a very similar process.

Design thinking differs from many traditional problem-solving methods. Many of these methods, such as traditional logical problem-solving, are very linear and follow a fixed set of steps from start to finish. In contrast, design thinking is very iterative, moving back and forth among the steps as often as needed. For example, it is very common for prototyping to lead back to the ideation or even the empathize phases to incorporate an insight gained from testing potential solutions with the stakeholders.

Another major difference is the strong emphasis on beginning with a deep understanding of the problem from the perspective of the people who face the problem in real life. The empathize phase often involves including those people in generating and testing solutions. Many design teams consist of people outside of the community or organization with special expertise in design-thinking methods working collaboratively with the people in the community or organization. The eventual goal of design-thinking leaders, however, is to disseminate design-thinking skills so widely that anyone can begin to apply these skills to their own problems and experiences. Even if you personally experience the problem you're trying to solve, however, you still need to consult widely in the empathize phase to make sure you're drawing insights from the thoughts and experiences of many people. Also, designers know eliciting fresh, outside perspectives during the idea-generating phase produces more potential solutions and higher-quality solutions.

Traditional logical problem-solving methods often work well for defined, structured, and well-understood problems. Troubleshooting a defective electrical circuit in a machine is a great example. Design thinking excels when people encounter new and messy problems in uncertain situations. Redesigning a household appliance such as a microwave oven for people who are blind is a problem like this. New, creative, and unexpected solutions are the hallmark of a design-thinking approach.

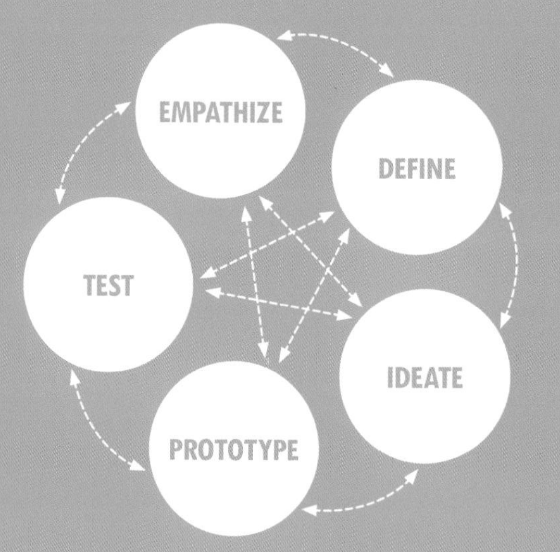

Design thinking is not a simple, linear progression. Because it is centered on the needs of people, it generally begins with the empathize phase. As a creative process, it then flows fluidly from any one step to another as needed. For example, you could move from the prototype phase to the define phase if a prototype reveals an unanticipated variable in how people use the invention. Like improvisational jazz, you move with the flow to create the best solution.

He found two main alternatives to sign language. Many people try to have paper and pen handy when they go out. Others try to type on the screens of their cell phones. Neither alternative works well enough in all circumstances. Message pads get left home or buried in purses or backpacks. Typing on a small keyboard under pressure can frustrate anyone—and bring more frustration than clarity. Texting a message with abbreviations and other shorthand might work with a person you already know, but not necessarily with strangers.

Effective communication with nonsigning people seemed to rank high among the needs of people with speech difficulties. We need to interact with people we do not know many times in a day. Grocery stores, airports, sports events, or auto repair shops can all present high-stress encounters. While the need for better communication was clear, the solution was not.

When he looked back on this project later, he said he would have talked to people with speech difficulties at this stage, rather than rely on secondary sources. "I would start right off the bat with interviews to learn more directly from people with speech difficulties," he said. "I think I wanted to get into the technical stuff right away." Also, "I was just shy," he admitted.

The Next Steps

Rather than improving old solutions, Austin wanted to explore new approaches. He knew people too often stick with one idea and never imagine better, more creative possibilities. Austin wrote a problem statement specifying what people with speech difficulties need to accomplish: to communicate with strangers who don't know sign language. Following design-thinking principles, he didn't decide beforehand what the solution would be. In design thinking, a needs statement focuses on actions (need to do), rather than specific things (need this particular tool). Using this approach, Austin left the solution open to a wide range of potential alternatives to explore, rather than fixating on only one.

With this problem statement in hand, Austin then generated as many ideas as he could, both on his own and with classmates.

A handier carrier for pen and paper? Maybe preprinted cards with common messages? A community effort to promote basic sign language proficiency? A portable speech synthesizer? As with classic brainstorming methods, he added wild ideas to the list, knowing that sometimes what sounds impossible can lead to practical solutions.

The Glove Idea

One day, one of Austin's gamer friends told him about a new gaming glove. The Peregrine is a wearable game controller, a button pad built into a glove. Touching different fingers and combinations of fingers executed different commands. This reminded Austin of the Marvel superhero Iron Man, who uses tech-powered gloves to fight villains.

Speech and Language Disorders

According to the National Institutes of Health, between six and eight million people in the United States have a language disorder. Speech disorder, one type of language disorder, is a surprisingly complex topic with many different forms and causes. For example, people with hearing disorders often have a speech disorder.

Many organs work together to produce speech. Difficulties with any of them may cause a speech disorder. The larynx, the tongue, the mouth, and the speech centers in the brain are a few of the main ones.

The National Institute on Deafness and Other Communication Disorders is a good place to learn more and to find resources for particular types of speech disorder. Visit their website at https://www.nidcd.nih.gov/.

Austin loved the adventures of Iron Man and his human alter ego, Tony Stark, a brilliant inventor and technologist. Coincidence? Probably not. Around this same time, one of the images of the red-and-gold–clad superhero and his luminous gloves started Austin thinking, *What about a device you could wear like a smartwatch to help communicate with other people?*

Austin found that other inventors had tried to make a sign language glove, but no one could get a signing glove to work. All the other prototypes were too big, too complex, and needed too much computing power. "That gave me a jumping off point," Austin said. "I researched what people had tried to do already. Then I tried to figure out, okay, what hasn't been done that I can reasonably attempt to do?"

As he sketched out various ideas, Austin saw a potential solution. He decided on a simpler, less complex approach. He pictured a glove device based on the Peregrine gloves with a readout screen. This would allow people with speech difficulties to "type out" their messages with hand gestures. Using this device, they could easily converse with those who do not understand sign language. "Since it's much simpler than a full sign language glove, I might actually be able to create it," Austin realized. He thought he had found a match between a human need and a technical possibility. He could create this "talking glove" and make a difference in the lives of people with speech difficulties.

Austin was eager to get started. Using his earlier experiences with inventing and his current design-thinking class, he planned his project. Revising his earlier sketches gave form to his vision. What could this glove that would turn finger movements into text display look like?

He also listed what he needed to do to make his glove device. He checked off what he already knew how to do, leaving what he needed to learn. Austin quickly saw he did not yet have all the skills he needed to complete this project. Both computer programming and microelectronics skills were high on the lists of both "need to know" and "do not yet know."

Learning to Code

There was a lot to learn, Austin knew. Learning either programming or microelectronics could be a big job. Learning *both* was intimidating. He wondered how long it would take him to gain the skills to make his idea a reality. He knew he could do it if he worked hard and long enough. But did he have enough time for this project and still have room in his life for friends, school, and sports? Would it make more sense to pick an easier project to fit an already busy schedule?

But the glove intrigued Austin. The image of the frustrated shopper stuck with him. It would also be a good inventing challenge and matched well with his design-thinking course. Even if programming and microelectronics were hard and took a long time to learn, they were valuable skills to have in his tool kit. *Why not get started now on skills I would need one day anyway,* he thought. Also, it wasn't as if he needed to be a professional programmer or an electrical engineer for this project. Austin realized he could get started and then learn more along the way. He began his first prototype, which provided both drive and direction for his studies.

Austin's school offered both an introductory and a more advanced programming class. The introductory course would be more manageable, but Austin knew he needed intermediate programming. He signed up for the advanced class to get his project moving quicker. At first, he found the programming difficult. *What did I get myself into?* he sometimes wondered. Though reserved, Austin was also open and approachable. He easily made friends in the class, and his friends helped him with tricky problems and shared programming tips. By working hard and getting the support he needed, he made rapid progress. He told his teacher about his project and, appreciating his determination, she gave him extra help outside of class. Before long, Austin knew enough programming to launch his invention project. He also knew he'd need to keep learning along the way.

Learning Electronics

Developing Austin's electronic skills turned out to be even harder than learning to program. Austin could not find good introductory material. Most of what Austin came across fell into two categories. Many books sketched out complex projects for those who were already skilled in microelectronics. Other books described very simple projects such as making an electromagnet. *Why is it so hard to find good basic information?* he wondered. Learning electronics felt like a gigantic puzzle, and he would have to crack the code.

With this insight, Austin started his own do-it-yourself learning project. "I had an idea of where I needed to be and then I worked backwards from that. You read an article and you figure out what doesn't make sense. You click a link and find out what that means. At the start it's a lot of just following chains of links. You don't understand anything at all, but you're getting closer to more fundamental concepts. Then you start learning the fundamental stuff and build your way back up slowly," he said. "I had a lot of browser tabs open at once.

"I mostly spent the first year with online guides and forums and YouTube lessons just learning how all this different stuff worked," he recalled. Austin also haunted electronics websites, such as Adafruit, SparkFun, and the Arduino company's website, and read everything. These sites provided some important basic pieces of the puzzle. Next, he bought parts that seemed helpful and read the instructions provided by the manufacturer. Sometimes he ended up with items he didn't need. Sometimes even the right piece was hard to figure out. He read and reread sections until they made sense. Or mostly made sense. "I was really missing someone I could just ask when I had a question," he said. Often he found that wiring parts and trying things out until it became clear was the way forward.

He worked backward from information about a specific electronic part to the bigger picture linking the parts together. Using this approach, Austin taught himself the fundamentals of electronics and the know-how he needed. As he moved along, Austin also found he needed to know more about how electricity worked. Sometimes he jumped ahead with a breakthrough. Sometimes he needed to step back to learn more, especially after "making stupid, damaging mistakes to my components." But Austin learned what he needed to know by working on his first prototype. His project was his teacher, though sometimes the lessons were tough.

Some days this task seemed endless. Why was this taking so long? When would he know enough to complete his invention? Both programming and electronics were subjects you could study forever and still not know everything. But having a clear vision of his talking glove gave Austin a specific target, so he continued to make progress.

"One year," Austin said. "I didn't think it would go past one year and that would be it. Then I got to the end of the first year and I'd spent most of it teaching myself electrical engineering and programming to be able to make anything. By then, I actually had made something, and I was like, 'Oh, this is really cool. I want to continue it.' So, I did it for another year. Then I continued for one more year after that. By the end of junior year, I was really surprised that it had gone on for three years. It was never the intention, but it worked out that way. I just cannot seem to judge how long something will take me to do."

The Invention: Touch to Talk

After those three years of hard work, Austin developed a device he called Touch to Talk. Touch to Talk is a glove loaded with electronics so people with speech difficulties can communicate with those who do not sign. People need not plan ahead of time to talk together. The person wearing the glove simply makes hand gestures that the device interprets as letters or numbers. Touch to Talk translates hand gestures into text. A large smartwatch-sized screen on the wrist displays the text to a conversation partner.

The "brain" of Touch to Talk is an Arduino Pro Mini microcontroller. Microcontrollers are small, single-purpose computers on a chip.

Pictured here plugged into a breadboard, the Arduino Pro Mini can store up to 16 kilobytes of code. In addition to devices like Austin's glove, the Pro Mini is used in robots, vehicles, home appliances, and games.

Widely used in modern electronics, these smart chips run almost everything connected to power. Smartphones, microwave ovens, cars, and solar panels all use chips. Their small size, low cost, and low power use allows designers to make electronic devices "intelligent." Smartphones make calls at the push of a button. Microwave ovens turn off automatically at the right time. Home sensors turn lights and heat on and off on schedule. Digital video recorders save selected television programs.

The Pro Mini is a favorite among hobbyists and designers. Created for ease of use, it has a simple, open-source circuit layout. Highly adaptable, these chips work in a wide variety of settings. This flexibility allowed Austin to be creative. The microcontroller's ease of use was also an enormous advantage since Austin was still new to electronics and to programming.

The core of Austin's invention is the Peregrine glove. Austin reprogrammed the gaming glove for alphanumeric characters instead of game commands. Stainless steel strips woven into the fingers complete an electrical circuit when your fingers touch. The microcontroller interprets different finger combinations as different letters or numbers. For example, touch the tip of your thumb to the tip of your pointer finger to make the familiar "OK" sign. This action would produce a keystroke, say the letter *A*. Your hand becomes a keyboard.

Not only can Touch to Talk spell out numbers and letters, but it also uses gestures as entire phrases. For example, a single hand position, instead of representing the letter *A*, might stand for "Thank you" or "What time is it?" For this reason, Austin's glove has more "keys" than a typical keyboard.

The final major part of Austin's Touch to Talk device is the display screen. Attached to the back of the glove, near the wrist, the screen is about the size of a smartwatch. The user manipulates the gaming glove to display the message. Finger movements become a text message. The "speaker" shares this text message with a conversation partner without needing to use their voice.

For his display, Austin used the screen of an old Nokia cell phone. Widely used, the Nokia 5110 liquid crystal display (LCD) screen is readily available, and its technical details are easy to find online. It is also easy to connect with the microcontroller, so it fit well into Austin's design.

Austin used parts of existing devices rather than create everything from scratch. The Peregrine glove, the Arduino Pro Mini, and the Nokia 5110 LCD screen were all on the market. The Peregrine glove used the finger-touch design Austin envisioned. Using the Peregrine, Austin saved both time and money. The explosion in do-it-yourself (DIY) electronics supports the creativity and ingenuity of hobbyists, inventors, artists, and makers. Many electronic components are available and affordable for inventors like Austin and for anyone who wants to learn electrical engineering.

Putting It Together

After finishing his plan, Austin built a prototype on a breadboard. This thin plastic base with holes to plug in and connect electronic components without soldering lets designers check their ideas and make changes easily. By assembling the components on a breadboard, Austin could see the entire circuit better and check his idea.

As he was working on circuit design, Austin was also applying his new coding skills to program the Pro Mini. Writing code requires using the programming language precisely and logically. The computer is stricter about grammar, syntax, and punctuation than any high school English teacher. Austin made coding mistakes. He fixed them. He learned.

After months of work, one afternoon, when Austin moved his fingers in the glove, the screen lit up with the letters he had just signaled. *Bingo!* He had proven that his concept worked. Digital messages could flow from gestures rather than spoken words. "Getting the hardware working like that, the first time, it felt like a huge step forward," Austin said. "I don't know exactly how to describe it. It's still the best feeling in engineering in my opinion . . . just making things work . . . it feels like bringing something into the world." Reflecting on that moment, Austin recalled, "Especially being young, it's a moment where you learn. . . . I learned that I could do more than I thought I could do. It's a complete endorphin rush. I did it! It's here! It works! Yeah!"

Touch to Talk was so much more than a good idea. Austin shared the news with his parents and friends, who delighted in his success. This was a major step. Just getting these parts to work together is a major achievement in designing electronics. But even as he celebrated this milestone, Austin knew there was much more to do. This working prototype was just one step of many in completing a finished product. For people to use it in daily life, Touch to Talk needed to fit comfortably on a hand and wrist. It also needed to stand up to the normal conditions and stresses of daily life.

Revising and Improving

Before Austin could consider his project complete, he had to solve two major problems in usefulness and practicality: size and energy use. The device was too big and would be too awkward for someone to wear. Austin knew his device had to be very portable. People carried devices such as smartphones and tablet computers in a backpack or purse, and if Touch to Talk were that size, someone would need to pull it from a pocket or a bag to use. This would not help much for immediate, unplanned conversations. Besides, would people carry *two* smartphone-sized devices? No, a device worn on the wrist was more sensible. Touch to Talk would have to get smaller.

Austin had to pack his invention into a wrist-sized package. He knew the circuit board needed to be smaller, but a smaller board couldn't perform all he needed it to. How to shrink it? What if he ground the edges off the circuit board to make it smaller and better able to fit? Austin knew about a computer-controlled machine that shaved off excess material. Austin and several friends built one of these milling machines for their school's innovation lab. Using it, Austin shaved a few millimeters from the Arduino board. His device became smaller and a better fit on a human hand.

The second big problem was that the device used too much energy. The prototype struggled to run using the battery, and the battery drained far too quickly. A bigger battery, however, made the device even more awkward to wear on the wrist. Frequent daily charging or battery changes were annoying and inconvenient. Austin puzzled over how to reduce power use as he worked on reducing size. He considered several ways to save energy. He experimented with different batteries and charging systems. Was there a smaller battery that could sustain the power required? Was there a simple easier way to recharge? Austin rewrote his code to use less power in the microcontroller. This helped some, but it was not enough. He kept coming back to the display, the primary use of power.

The Nokia 5110 backlit LCD display in his first prototype needed constant power. This constant draw burned a lot of energy. He thought about how people would use Touch to Talk. Running out of energy on a device always needed at a moment's notice was a huge drawback. So, he looked at other types of displays that drew less power.

The first alternative Austin explored was an e-ink display, like those that the Kindle and other e-readers use. These displays, looking like ink on paper, are less clear than the LCD, but they are bright and have high contrast. They are also readable in direct sunlight and across a wide viewing range. Readers can see these displays from the sides and straight down from the top. Best of all for Austin, an e-ink display needs little power to display the message. Unlike the LCD, only message updates required much power. Holding messages was much more important for users than allowing a bit more fuzziness of the image, Austin explains. Engineers need to make these trade-offs all the time.

Austin tested the e-ink display on his second prototype. "Surprisingly, the readability suffered too much," he said. "So, I had to try a third option. My final prototype uses an OLED [organic light-emitting diode] display." Austin's third version ran on much less energy. His glove could operate for a full month of moderate daily use from a small battery. This worked well for everyday use.

Eliminating Cross Talk

After solving these issues, new problems arose. Touch to Talk malfunctioned at odd and unpredictable times. It turned off when it shouldn't or didn't turn on when it should. Or it sent the wrong signal. And it got too hot. Since inventing requires solving problem after problem, Austin tackled these additional issues with quiet resolve. He knew cross talk was the cause.

Electricity and magnetism are two parts of the same force of nature. When electricity flows through a wire, the current creates a magnetic field around the wire. That is why the needle on a compass will move when put near a circuit. In turn, passing a wire through a magnetic field creates (or *induces*) an electric current in that wire.

Some flashlights with no batteries work by sliding a powerful magnet past coiled wires. The induction current created by the sliding magnet lights the bulb. This induction effect makes electrical motors and generators work. When induction causes a circuit to stop working, however, we call it cross talk.

All the closely packed wiring and rapidly changing signals in Austin's glove triggered cross talk. Working out cross talk glitches was a tedious troubleshooting process. Austin needed to analyze each circuit, try a fix, and test to see if the fix worked. Then he could move on to the next circuit until everything ran correctly.

Science Fairs

In his sophomore year in high school, Austin submitted Touch to Talk to the San Mateo County Science Fair. "It was really fun, really casual," he recalled. "People milled around and asked you questions . . . everyone was just having a good time presenting their work . . . and talking with their neighbors." He tied for first place with a classmate in the engineering category.

From there, he went on to the San Francisco Bay Area Science Fair. "It was a lot more competitive. . . . You felt the tension in the air," Austin said. Adding to the tension, cross talk issues bedeviled the glove at inopportune times. He was still feverishly troubleshooting cross talk glitches as the judges were coming to view his project. Yet even with these technical issues, Touch to Talk received first place in the embedded systems category and the grand prize at the fair.

This meant that Austin qualified for the ISEF and a trip to Pittsburgh, Pennsylvania, site of that year's fair. "It was a really incredible experience," Austin said. "It was fun to meet everyone at ISEF. . . . I felt very supported by my teachers, especially Amanda Alonzo, my biology teacher . . . she was my mentor." Thinking back on the experience, he added, "It's also great because you get there and you're meeting people from all over the world. . . . I realized that there are leagues higher than where I am right now in terms of my skills. So, it was definitely humbling."

But along with all the fun and excitement, Austin also had to cope with technical glitches. The cross talk issues came back again. "When I was at ISEF, I was up every night working on the glove to repair it," he remembered. "When it came to the day of the event, it still wasn't working completely during the judging." Even though the glove's performance that day was disappointing, he could show that the design worked. "I had two versions of the glove. I had the built-up glove . . . and I had a bench top version . . . my test version." His bench version, although not fully assembled and still connected to a test board, proved the concept worked, even if the final version of the glove was not functioning. The judges commended his work and his invention, but Austin did not win in his category. Regardless, Austin said, "The real fun was just in being there. I felt really lucky to be there."

Following ISEF, he plunged back into the whirlwind of his senior year. Touch to Talk was Austin's senior project, and he incorporated the lessons he'd learned at the science fairs. Along with schoolwork and sports, Austin spent several hours each week on inventing. He invested over one thousand hours developing Touch to Talk. The night when the senior class displayed their projects, Austin's table with Touch to Talk was a magnet for both the socially conscious and the technically inclined. And in his high school, firmly planted in Silicon Valley, many of the parents, relatives, and friends attending that night worked in the tech industry or were very tech savvy, so the praise he received represented significant merit.

The Future for Touch to Talk

Getting Touch to Talk to work as he designed it was a major milestone for Austin. Touch to Talk needed testing by people with speech difficulties under normal, everyday conditions. Their feedback and suggestions were necessary for continued improvement. Setting up this testing, however, was harder than Austin expected. One promising lead fell through just when he was ready to start. Many questions still needed answering.

Was the device light enough and comfortable enough to wear all day? Was the glove too hot to wear for long periods? Did it hold up to the bumps and scrapes of a busy day? Was it easy enough and quick enough for frequent use? Was the battery charging easy and convenient? What maintenance did the glove and the electronics need? These were questions that only real-life users who have speech difficulties can answer. "I can't speak for their experience," Austin said, "but I would love to hear their perspective."

Austin also considered adapting Touch to Talk for American Sign Language (ASL). Adding sensors to track motion would make this possible. With this capability, clients who sign would not need to learn unfamiliar gestures to type out messages. They could use Touch to Talk right away. But although chip technology continued to advance rapidly, there were still major technical hurdles to overcome before Austin could create an ASL glove. "It would require very different hardware than I had. My approach was meant for a simple and cost-effective method," Austin said. "As far as I know, no one has succeeded yet in creating a marketable ASL glove because they are so complicated and processing power intensive. You need to have a computer nearby to handle the interpretation."

Further advances in chip processing power and in battery capacity technology would be needed "for now, at least," Austin said. "But I look forward to when that's not the case anymore. I think if someone came to me and said, 'I really need this,' then that would motivate me. Of all the reasons [to continue], that would be it."

Austin, like most of his classmates, planned on attending college right after high school. But what college would be best for a seasoned inventor? From his teachers, Austin heard about Olin College of Engineering, a new and innovative school in Needham, Massachusetts. Olin aims to develop engineers to focus on human needs and solve human problems. In most Olin classes, students make and build, rather than just read and discuss. This seemed like a great match for Austin's interest in designing and inventing. He applied, and Olin accepted him.

Before going to college, however, Austin took a year to work with AmeriCorps. Members of AmeriCorps work on community projects all over the country. They help in schools, maintain national parks, repair public buildings, and help neighborhoods.

His experiences in AmeriCorps changed him. He refocused his goals and wanted to center his life's work on designing tech for social benefit, rather than simply taking any high-paying, high-tech job. "Most of that change came from the people around me. . . . In AmeriCorps people were coming from all over the country from a lot of different backgrounds," Austin said. "They had perspectives that were very different from mine coming from Silicon Valley."

While in AmeriCorps, he saw "different problems that [I] hadn't been exposed to prior to that—more serious, more human, more dire problems." His experience in AmeriCorps, he said, "gave me much more of a shift in perspective, much more of a shift in perspective than I would have had if I had gone straight from Nueva [his high school in affluent San Mateo, California] to Olin."

Austin and his AmeriCorps friends discussed how venture capitalists invest their funds. "It's so easy, relatively speaking, to raise money for interesting tech ventures. In comparison, it's so much harder to raise that same money for solving problems that really exist on this planet right now in terms of food distribution or providing shelter for everyone." These experiences led Austin to consider going to a different college. His Olin advisers persuaded him that the social problems he wanted to work on and the organizations that tackle them need people with the technical and engineering skills he would continue to develop at Olin. After his life-changing experience in AmeriCorps, Austin came to the Olin campus with a new sense of purpose.

Beginning his studies at Olin, Austin felt, was "definitely the right choice for me. The learning style really works for me because it's very hands-on and project-based. The professors give their classes around where people are struggling at the moment. They are there to help you learn to do it and not to lecture. You end up doing stuff you didn't think you could do at all."

In the spring of 2020, the COVID-19 pandemic swept the globe. Olin, like nearly all US colleges, closed their campus and sent students home to shelter in place with their families. During this time, Austin's mother became a volunteer for the Community Emergency Response Team (CERT), helping their small California coastal county prepare to help those in need during the pandemic shutdown. She arranged workshops about what to do in emergencies, such as an earthquake, during the pandemic, created teams in each neighborhood to check on the especially vulnerable, and set up phone trees to keep people in contact with one another. Austin worked with her and the local fire chief to write and distribute a weekly newsletter to over one thousand residents with the latest information on resources available in case of emergency.

He also worked "with a nonprofit in the area that provides services for the Latinx community. They do cultural arts programs for kids, as well as mental health and counseling services . . . they do a lot of outreach and support for families during [the pandemic]," Austin said. He helped revamp their whole IT backend so they could expand their services, and worked on creating a new website for them.

Like many students during the pandemic shutdown, Austin attended his college remotely. "The most recent human-centered design project I worked on was through a class." He and his partner worked with a man who was hard of hearing. "His wife's concern was that she might be gone, a fire would start, and he wouldn't wake up in time to get out." Austin explained that they worked on adapting existing sound-based alarms, such as smoke alarms and carbon monoxide alarms, to use flashing lights or vibration. "It was a good project," he said.

Whether online or remote, Austin planned to finish his studies at Olin. But after that, he wasn't yet sure where the future would take him. "I want a similar experience [to AmeriCorps], something radical that gives me a totally new perspective. Beyond that I want to work for nonprofits. . . . There are not enough people providing technological support for nonprofits and I want to be one of them."

Sometimes in life, however, the truly tragic happens. Not long after Austin and I talked about his future in this final interview, he died unexpectedly.

There are never any satisfactory explanations for the sudden absence of someone so alive, so vital, and so full of promise. I was in shock at hearing of his death. Austin was kind, compassionate, brilliant, and unassuming, leading by example and through his deeds. His family and his many friends, including myself, miss him greatly.

Through hard work, skill, and care, Austin made the world a better place. His legacy lives on through the next generation of engineers, design thinkers, and socially conscious leaders who use their talents to solve real problems for real people.

5

DEEPIKA KURUP AND PURIFYING WATER THROUGH SUNLIGHT AND SCIENCE

Deepika Kurup learned about global issues early in life. Every summer she and her family traveled from their Nashua, New Hampshire, home to India to visit family and friends. "I've always enjoyed going. It's a beautiful place—there are palm trees, beaches, and so many sights, sounds, and colors," Deepika says.

The summer weather in India is hot and humid. Deepika and her sister drank plenty of water to stay hydrated. During these summer visits, her parents insisted that they drink only boiled or bottled water. "Unlike here in America, where I can just turn on a tap and easily get clean, potable water," Deepika says, "in India the water is often contaminated."

While her parents ensured that their family had clean water, Deepika learned that not everyone in India had safe water to drink. "I saw people who were drinking extremely dirty water on the side of the road. It was the same water they used to wash clothes and bathe in." She watched people standing in long lines to fill containers from a public tap. This firsthand experience with people who had no access to clean, safe water opened her eyes. "That's what made me want to make a difference in this area." Deepika recalls. "Watching these kids being forced to drink this water compelled me to address the water crisis."

A Quick Perspective on the Global Water Situation

Why do so many people have to drink dirty, contaminated water? Although our planet has vast oceans and water covers 71 percent of Earth's surface, only about 2.5 percent of that water is fresh water. Of that 2.5 percent, Deepika says, "less than 1 percent of Earth's freshwater supply is available for human consumption."

In many places, like much of India, getting clean water is very difficult. Inadequate sanitation facilities, agricultural runoff, and industrial waste pollute much of the water. Harmful bacteria and pathogens, organic chemicals, and toxic heavy metals, such as lead, can all contaminate drinking water. UNICEF estimates that three thousand children die each day from lack of clean drinking water. The World Bank reports that 1.5 billion people around the world suffer from water-related diseases such as cholera, dysentery, diarrhea, typhoid, and polio each year. Safe, affordable, sustainable, and easily deployable water purification techniques could save 3.4 million lives each year, according to the World Health Organization (WHO). The WHO also reports that one-third of the world's population, or 2.5 billion people, do not have safely managed drinking water in their homes. Over 785 million people—about 10 percent of the global population—must travel up to half an hour to get clean water. Of these, 80 percent live in rural areas in the developing world.

A study published by the World Bank links poverty with scarcity of clean water. The two often go together. In developing areas, women and girls spend a total of two hundred million hours *every day* collecting water. That is time not spent on caring for children, working at income-producing jobs, raising food, or getting an education. They could use this time to escape the cycle of poverty that traps families for generations.

"With rising populations, industrial development, and economic growth, our demand for clean water is increasing," says Deepika, "yet our freshwater resources are rapidly depleting." The WHO estimates that by 2025, if trends continue, half the world's population will live in areas lacking enough clean water.

Decision to Act

"The global water crisis is something that's close to my heart as I was affected by it personally when I was seeing these people who don't have access to clean water, something we all need to survive," says Deepika.

That summer she acted. "I decided that I wanted to combine my passion for solving the global water crisis with my interest in science." To start, she says, "I did what most people do when they see something they don't understand. I googled the questions I had." She learned the hard facts about access to clean water from UNICEF, the World Bank, and the WHO, giving context to her firsthand experience with the water crisis.

"That's when I realized science could be applied to modern problems today," Deepika says. "It helped me make a connection between the scientists you learn about in school who created vaccines in the early eighteenth century and the researchers who are working right now to solve exciting problems."

Google to Garage

Deepika was interested in science from an early age. "When I was younger, I always loved to build things," Deepika recalls. "I can vividly remember helping my dad assemble bookshelves and bicycles and building motorized windmills out of K'Nex and Legos." Perpetually curious, Deepika says, "Science always had an explanation for all my questions." She believes that STEM education has the power to lead us to revolutionary new discoveries that can solve global challenges.

She found, however, that not everyone shared her enthusiasm for science. Most of her peers at school focused on sports, socializing, and drama. "Not many students paid attention to a girl interested in STEM." Deepika tried to involve others in STEM by starting a science bowl team, but "only five kids showed interest." She couldn't help but contrast this to the high turnout for sports teams and cheerleading.

"Recently I've been seeing a change in this attitude," Deepika notes. "Kids like me are slowly starting to receive more recognition and opportunities for pursuing our passion in science." Given these early experiences, it's easy to see why Deepika immediately looked to science for a solution to the problem she observed firsthand.

First Steps

Once resolved to help solve the problem of getting clean drinking water to rural India, she started her research. "I googled, 'Why don't people have access to clean drinking water?' and 'What can be done to help these people?'" she says. And was anyone already leading any efforts to fix this situation?

She soon found there were many methods of purifying water. Cities purify water by filtering it, treating it with chemicals, infusing it with ozone, aerating it, shining ultraviolet (UV) radiation on it, or forcing the water through a membrane with tiny pores.

In the US, cities have large, well-equipped water treatment facilities to purify water. Deepika saw that these methods would not work well for rural India, which has very little infrastructure for water and sanitation. Rural India does not have much electricity or expensive equipment. Slow, physical separation processes to purify water took too long and didn't produce enough drinkable water. Some chemical processes left the water with an aftertaste or a foul smell. And some compounds resulting from chemical treatment were pollutants.

She realized that India needed a simple yet effective water treatment method. "There is a pressing need for green, sustainable, easy-to-use, inexpensive, and effective technologies for water purification."

Appropriate Technology

Appropriate technology is a philosophy, a movement, and a set of design principles arguing that technology and its products should fit the social, cultural, and economic environment of the local area. Technology should help people manage their lives better and control their own futures. Appropriate technology is green: energy efficient, nonpolluting, and sustainable.

Designers and inventors applying appropriate technology use low-cost materials to produce high-impact solutions for local problems. They need creativity, persistence, and collaboration with the community. The original solar disinfection process shows these characteristics. Clear 2-liter (2.1 quart) soda bottles are readily available and reusable. The energy for disinfection comes from the sun, and it is easy to use at home.

William Kamkwamba's electricity generating windmill (see chapter 3) is another good example. He built his windmill from materials scavenged from local landfills and an old bicycle, along with a bicycle lamp generator. This windmill produces the energy to power William's home without polluting. Other examples include water pumps powered by bicycles, homes heated by solar energy, and clay pot refrigerators cooled by evaporating water.

Often people think appropriate technology is limited only to helping people in developing nations. Developed nations also use appropriate technology to save natural resources. For example, the National Center for Appropriate Technology (NCAT) helps communities save energy with appropriate technology for the home and workplace. NCAT shows farmers and homeowners how to use wind and solar power in ways that would be familiar to William. NCAT also promotes biofuels, geothermal energy, and microhydropower. Most hydropower comes from rushing rivers or waterfalls, such as Niagara Falls in the US and Canada or Nkula Falls in Malawi. Microhydropower harnesses the energy of local rivers to make electricity for individual farm or home use. NCAT has expanded its mission to include sustainable farming techniques along with clean, renewable energy.

Mahatma Gandhi, in addition to helping liberate India from British colonialism, was an early advocate for small technologies. He saw clearly that helping rural villages toward self-reliance was essential in supporting and sustaining local control of the country. E. F. Schumacher's 1973 book *Small Is Beautiful: A Study of Economics as If People Mattered* helped propel this movement into the national consciousness. Designers such as Buckminster Fuller, Amory Lovins, and Victor Papanek all contributed their ideas and applied these principles to their inventions. The d.school at Stanford is famous for inventions such as a solar-powered incubator that saves the lives of many infants and a low-cost, solar-powered lamp for replacing unhealthy kerosene lamps like the ones William Kamkwamba grew up with.

Amy Smith of MIT, founder of the MIT D-Lab, devotes her life to helping improve the lives of the "bottom billion." These billion people in developing nations live on a dollar a day or less. Smith and her students look for simple ways to make a large impact. For example, one project teaches people to make charcoal for cooking from waste sugarcane. Generations of families cutting down trees for cooking fuel has deforested many regions of the developing world. And many children die from respiratory disease caused by inhaling smoke from cooking fires. Cleaner-burning charcoal will help reduce that number. The charcoal made from sugarcane is cheap and burns cleanly. It will also save trees that would have otherwise been cut for fuel. And people can earn money making charcoal for the community. Smith is also a strong advocate for women and people of color in engineering. "It would be great if students recognized that engineering with a humanitarian focus is as legitimate as aerospace and automotive engineering," she says.

Finding SODIS

Continuing her research, Deepika learned about a method of water purification using sunlight and plastic water bottles. Lebanese scientist Aftim Acra developed this method of solar disinfection in the early 1980s and called it SODIS for short. Swiss research groups verified the effectiveness of SODIS and promoted the technique. Deepika was excited to find out about SODIS since it fit her criteria for a simple, appropriate technology approach.

SODIS works because UV light and heat from the sun can kill bacteria. SODIS is also easy to use and takes advantage of warm, sunny climates. A person fills a clean, clear, 2-liter (2.1 quart) plastic soda bottle with water and puts it in the bright sun for a few hours to create safe, drinkable water. That's all there is to it. Shaking a partially filled bottle before putting it out in the sun adds oxygen to the water, which also helps destroy pathogens, according to the Centers for Disease Control and Prevention.

SODIS had many of the advantages that Deepika was looking for. Research evidence had established the effectiveness of SODIS in killing pathogens. It was cheap since the only equipment needed were empty plastic soda bottles. Sunlight was abundant and free. People found little change to the taste of the water. Storing the treated water in the same plastic bottles was convenient, and the narrow bottle necks helped prevent reinfection of the water.

As Deepika learned, however, the original SODIS method had a major drawback. It was slow. Under ideal conditions, solar disinfection could take as little as two hours. But anything that blocked that sun's rays made SODIS less effective and meant the water needed more time in the sun. If the day was partly cloudy, the water bottles needed to spend six to eight hours in the sun. If clouds covered more than half the sky, SODIS took two full days for the treatment to kill the pathogens.

And even with these guidelines, how could someone be sure when the water was clean and microorganisms destroyed? The intensity of sunlight differs with cloud cover and weather. For home use, there was no simple way to tell when the water was ready.

Original SODIS

The original SODIS technique works in two principal ways to kill harmful pathogens. First, the water sitting exposed to sunlight heats up to around 122°F (50°C). Over time this high heat warps the enzymes within the cells of the pathogens. When these enzymes lose their shape, they cannot perform their essential functions and the pathogens die.

Second, SODIS uses the sun's UV rays to disrupt the structure of the DNA in the cells of the polluting microorganisms. The shape of DNA is often described as like a spiral staircase. UV rays can break some of the "steps" in this "staircase" and fuse them together in wrong combinations. This distorts the shape of the DNA so it cannot replicate. The strands of DNA will no longer match with their correct partner. Using another common DNA analogy, the "key" no longer fits in the "lock." With the DNA machinery no longer working, the pathogen eventually dies.

These two processes working together kill pathogens to disinfect the water. Since SODIS works by exposing water to UV rays, anything that interferes with light reaching the water slows the process. So, the process won't work on very cloudy days, and it takes much longer on partly cloudy days. Either glass or plastic bottles work, but they must be clear and free of scratches so that enough UV light can get through.

Finding Photocatalytic SODIS

Looking further, Deepika read about researchers who began coating the inside of the plastic bottles with nanoparticles of titanium dioxide to speed up the SODIS water treatment. This innovation took advantage of a process called photocatalysis.

A catalyst is a material that speeds up a chemical reaction, and *photo-* means "light." So a photocatalyst can accelerate solar disinfection. When sunlight hits a photocatalyst like titanium dioxide in a bottle of water, it changes some water molecules into reactive oxygen species (ROS). ROS are types of oxygen molecules that are quick to react chemically, such as hydrogen peroxide. These oxygen groups damage the DNA of the pathogens in the water and block their ability to reproduce. Unable to reproduce, the pathogens die.

Elaborating on this, Deepika says, "These reactive oxygen species can remove bacteria and organics and a whole lot of contaminants from drinking water." Adding the photocatalysts cuts purification time and makes SODIS much more practical.

"But, unfortunately," Deepika says, "there are several disadvantages to the way photocatalytic SODIS is currently deployed." For one, the photocatalytic coating is on the inside of the plastic bottles. This means that "they're blocking some of the UV radiation and diminishing the efficiency of the process," she says. Because of this property of titanium dioxide, sunscreen lotions often use it and other similar materials to absorb UV light.

Since these photocatalysts don't stick well to plastic, they wash off in the water getting treated. Then "people end up drinking the catalyst," Deepika says. The photocatalysts are not harmful to people. But after each use, some of the coating washes off, so the next time the bottle gets filled, it takes a bit longer for the water to purify. How much longer is difficult to tell, and it's hard for people to track these safety margins. "It's really inefficient if you keep drinking the catalyst," Deepika explains, "because then you have to continue to replenish it, even after a few uses."

Even though photocatalytic SODIS was faster than the original

SODIS method, Deepika saw that it was still too slow. She set out to accelerate the photocatalytic SODIS procedure.

The first part of the puzzle she tackled was preventing the coating from blocking the sun. How could she keep the photocatalyst in constant contact with both the solar UV radiation and the water? "I eventually thought of an idea for my own prototype," Deepika says. She decided to put a photocatalytic source *in* the bottle, rather than coating the inside of the bottle. Also, by having a photocatalytic object inside the bottle, more water would receive the catalytic effect sooner. She designed a prototype water bottle and inserted a rod of PVC pipe that she coated with a photocatalyst. Then nothing was blocking the sunlight from the water.

Improving the Photocatalyst Formula

What about the photocatalyst itself? Could she design a better, faster-acting mixture? Deepika continued her internet research and kept plowing through complex chemistry and physics articles in academic journals. Building on her hours of research, "I hypothesized that this disinfection process can be accelerated by using two photocatalysts: titanium dioxide (TiO_2) and zinc oxide (ZnO)."

To see if this combination improved results, she ran some comparative tests. "I didn't have any fancy laboratory, so I did my experiments in the kitchen," she says, "until my parents told me to move to the garage." Since she was testing polluted water, she admits, "I guess it wasn't very sanitary."

She set up her lab in the garage and continued her experiments. Taking three sections of PVC pipe, she coated one with just titanium dioxide, one with zinc oxide only, and one with both titanium dioxide and zinc oxide.

"To test my invention," she says, "I conducted experiments on samples of water taken from the Merrimack River." The Merrimack is "one of the most polluted waterways in New England," according to the Merrimack River Watershed Council, so it made for a tough test of her idea.

"Four bottles filled with this water were tested," she says. She put each of the three coated rods inside three different tall cylindrical bottles. A fourth identical bottle had no catalytic rod. This bottle served as an experimental control to compare to the others. At the bottom of each of the bottles, she put a parabolic reflector, "in order to concentrate solar energy along the entire length of the bottle." She put all four of the bottles into bright, direct sunlight.

How would she know if her idea worked? "I took a sample of water from each bottle at three-hour intervals and smeared them onto petri dishes with agar. I let the bacteria grow for seventy-two hours inside an incubator I made." This way she could track how much of the bacteria were killed by the sunlight in each time interval. Biologists often use the petri dish and agar method for growing bacteria in the lab. Petri dishes are shallow glass or plastic containers, roughly the size of a wide-mouth jar lid. Biologists fill petri dishes with a nutrient gel called agar that bacteria thrive on. This gives the bacteria both food and a good growing surface.

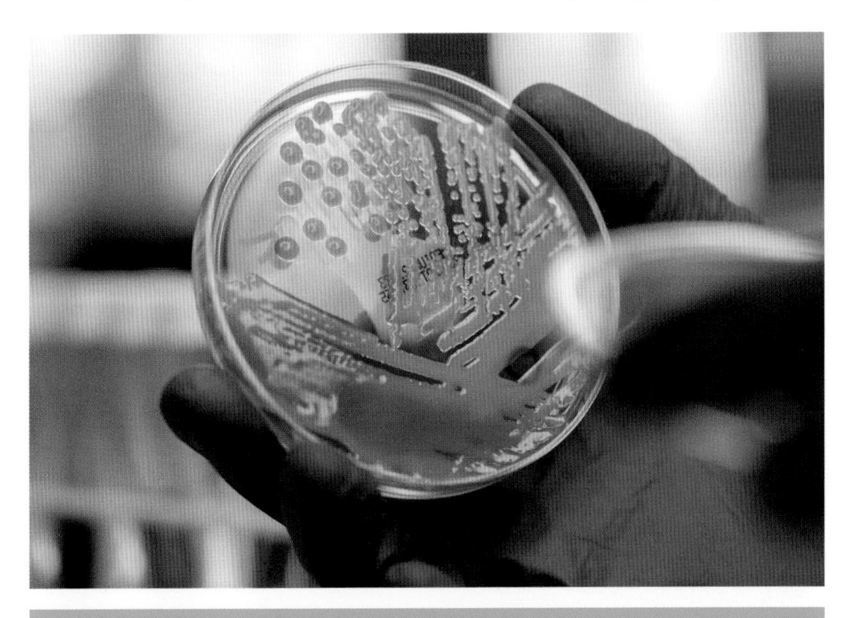

Petri dishes are most commonly used to study bacteria, but scientists can also grow other types of cells, such as nerve cells, to study and to experiment on them.

Deepika's petri dishes came out of the incubator, a special low-temperature oven, covered with colonies of bacteria. Then she counted the number of colonies on the petri dishes from each of the water samples. She used multiple dishes for each sample to average out any variations. The more colonies, the more contaminated the water.

She found the control bottle, with no photocatalytic rod, had the greatest number of bacterial colonies, just as she expected. The bottle with the mixture of the two photocatalysts grew the fewest colonies. "The results of my experiments show that adding zinc oxide to titanium dioxide increases the photocatalytic effect and reduces the amount of bacterial colonies," she says. She created graphs of her data and saw that they displayed this outcome clearly. Then she had a better formula for a solar disinfection photocatalyst.

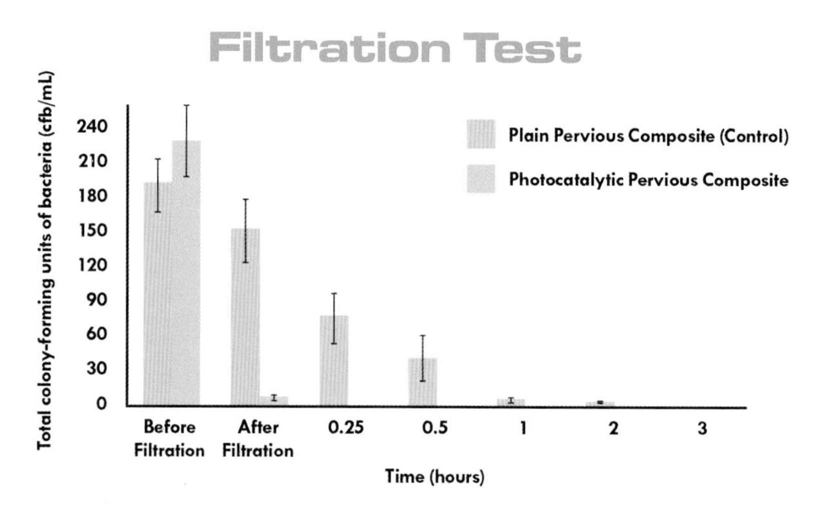

*Source: Graph reproduced from Deepika Kurup, "A Novel Photocatalytic Pervious Composite for Degrading Organics and Inactivating Bacteria in Wastewater" (entry paper for the Stockholm Junior Water Prize), 2014, https://www.wef.org/globalassets/assets-wef/3---resources/for-the-public/stockholm-junior-water-prize/winning-research/2.-u.s.-national-winners/2014-winning-paper---deepika-kurup.pdf.

This graph of Deepika's research data shows how the amount of bacteria in filtered water samples decreases over time when exposed to sunlight. One sample had only a pervious concrete filter, while the other had her photocatalytic concrete filter. The sample with the photocatalyst eliminated nearly all of the bacteria in a shorter time. Scientists like Deepika use graphs to summarize their results in one image.

These first experiments showed that her photocatalyst made SODIS work faster, but could she solve the problem of keeping the catalyst from washing off into the water? "My experiments did not always go as planned, and I had to look for alternative solutions," Deepika says. "For example, I initially struggled to find a proper binding agent for my photocatalytic composite."

In her initial tests, she used citric acid mixed with her chemicals to paint the PVC rods. Citric acid, found in lemon juice and added to many foods, would be safe in drinking water, but it would not hold very well. She tested many alternatives. Toxic adhesives held well but were not safe in drinking water. Nontoxic adhesives washed off too easily. "Although it was frustrating when my early methods did not work, I was determined not to get discouraged."

She finally found one safe product called EcoPoxy. EcoPoxy is a nontoxic and transparent glue that works well on a variety of surfaces, including plastic. Testing showed the EcoPoxy performed well on the surface of the PVC rods.

A Concrete Idea

As Deepika worked on her invention, her thinking continued to evolve. She had found a nontoxic binding, but would EcoPoxy be easily available in India and other developing nations? An adhesive that requires mixing in a specific ratio and application according to directions could complicate the process for local communities. And the cost of EcoPoxy—around $200 for 1.6 gallons (6 L)—added to the cost of the water bottle inserts. She wanted her solution to be simple to make and inexpensive, as well as effective. Was there a better way?

Then Deepika thought of a variation of her earlier idea to put the rod inside the bottle. Why not embed the photocatalyst in the rod itself instead of coating the surface? After some thought and testing, she realized that concrete would be an ideal material. Concrete was nontoxic, and she could add her photocatalysts directly into the mix of cement, sand, and water before the concrete set.

She laced concrete with the photocatalysts, then formed the gooey material into rods. Once the concrete solidified, she put her concrete rods inside her test SODIS bottles. The photocatalysts in the concrete wouldn't block the sun's UV rays and wouldn't wash off.

Again, she put these bottles in the sun after adding the contaminated Merrimack River water. This time, she also used some untreated water from the Nashua, New Hampshire, water treatment plant. Following her procedure, she cultured bacteria from the test bottles and from raw, untreated water from the river and the plant. After counting the bacterial colonies on her petri dishes, she found that her method was just as effective at purifying the water as the existing photocatalytic SODIS method.

But her combination of two photocatalysts worked faster than just one. And her method allowed for more sunlight to reach the photocatalysts. These two improvements significantly reduced the disinfection time, making it a potentially much better solution to the problem of cheaply and effectively cleaning contaminated water.

Entering the 3M Contest

Based on these encouraging results, Deepika submitted her improved SODIS method to the Discovery Education 3M Young Scientist Challenge. Her entry video showed her careful work and her rigorous scientific process, along with her obvious passion for this project. Chosen for the finals, Deepika spent her entire summer vacation refining her ideas, reading more articles in scientific journals, and consulting with her 3M mentor.

One problem she solved that summer was making the heavy concrete rods much lighter. The cement mixed with sand was very heavy. Working with her mentor, she discovered a product called "glass bubbles," which are microspheres of glass. She substituted these glass bubbles for the sand and got a strong, yet much lighter compound to make her photocatalytic rods. This made them easier to use inside the bottles and more practical.

After many rounds of testing for the best mix of her four materials, she came up with a ratio of one part zinc oxide, four parts titanium dioxide, twenty-five parts glass bubbles, and one hundred parts of cement, by weight. This formulation was both strong and very light.

The system she brought to the 3M finals could cheaply and easily purify water contaminated by harmful bacteria. She also tested this system with an organic chemical dye, methylene blue, to find out if her system could also help with removing harmful organic chemicals. Here, too, her results were promising, and her method reduced the amount of methylene blue in the test samples.

After a summer of concentrated work with her 3M mentor, Deepika was ready for the final event. She traveled to 3M headquarters in Saint Paul, where she met her mentor in person, along with the nine other Discovery Education 3M Young Scientist Challenge finalists.

Before the final judging, the 3M contest staff gave Deepika and the other finalists two on-the-spot challenges. The first challenge tested everyone's ability to think creatively in a short time. Contestants had to solve a problem posed by the judges using uncommon connections between different 3M products to design a novel solution. "We were also partnered up with one of the other contestants that we had just met the day before, so it was pretty intensive work," says Deepika. She and her partner sketched out an idea for a ceramic film to coat space exploration vehicles to protect them from micrometeor impacts that can pit the surface and increase drag. This film would be lightweight, inflammable, and easy to apply. The second challenge was to make a Rube Goldberg machine to turn on a light bulb. Simple enough—but the contestants needed to use their science and engineering skills to light the bulb in several steps with different forms of energy, such as chemical or mechanical energy.

On the last day, Deepika's presentation was clear and compelling. After explaining her first exposure to the global water crisis and seeing children her age forced to drink dirty water, she presented her device and results. Focused and unflappable, Deepika even kept going when she accidentally knocked over one of her props. She told her audience that her innovation might help up to "1.1 billion people in the world."

Her improvements to photocatalytic solar disinfection and her clear, brisk presentation won her the Discovery Education 3M Young Scientist Challenge award for 2012, along with a $25,000 prize. Deepika remembers this contest and the people she met fondly, and she encourages other teens to enter.

The White House and Obama

After winning the 3M contest prize, Deepika received several other honors, including the President's Environmental Youth Award.

"In late April, when I got a phone call from Discovery Education informing me that I was invited to the 2013 White House Science Fair, I was excited and overjoyed," she says.

At the White House, Deepika met other teens who were enthusiastic about solving world problems. For her, the opportunity to see the work of her peers and share ideas was one of the main benefits of this event. Deepika also met many scientists and celebrities, including Bill Nye, and talked with them about the impact of creative young scientists. "I realized that science is everywhere. Science has the enormous power to help people find solutions to problems we never thought could be solved," she says.

President Obama addressed the group and encouraged them to continue to dream, create, and innovate. "I truly believed him when he said that we are participants in this long line of inventors and creators that have made America the most dynamic economy and the most dynamic country on earth," Deepika recalls.

Stockholm Water Prize and Improvements

Deepika continued to work on her process and improve her methods. In 2014 she entered her invention in the Stockholm Junior Water Prize contest with several enhancements.

First, she added a filter. Another limitation of SODIS was that if the water was cloudy, then SODIS didn't work as well. Silt and organic matter suspended in the water both contaminated the water and blocked some of the UV rays. Filtering out these materials greatly improved the results.

Deepika thought about enhancements she could make to the filter. She remembered a technique that allowed cities to drain storm water from flat areas like parking lots quickly so they wouldn't flood. The builders used a mixture of concrete that was very pervious so water could flow through. They made this pervious concrete from mixing coarse sand and gravel with cement. She realized she could use pervious concrete as a filter. Deepika formed pervious concrete into disks to make the filters for these tests. Getting the right proportions for the pervious concrete took more experimenting.

She made the concrete filters pervious by altering the amount of coarse sand in the mixture. If the concrete was very dense with too little sand, the water dripped through very slowly. If too much sand was in the mixture, then the concrete was too porous. The water passed through quickly, and too much of the silt and organic matter flowed through as well. She spent many hours working out the best concrete mixture to get the right size for the pores.

Deepika also experimented to find the best size grain of sand, since sand particles come in many sizes from very fine to very coarse and large-grained. The coarser the sand, the bigger the pores in the concrete were and, thus, more pervious to the water. After much experimentation, she finally worked out a practical formula for her pervious concrete. Placed near the opening of the SODIS bottles, her disks successfully filtered out the organic matter as she poured water into the bottle.

Deepika realized that she could add the photocatalysts to her pervious concrete mixture, creating a more effective insert as well. A pervious insert would allow water to circulate through it, and the water would come into more contact with both the photocatalysts and create more ROS. This would speed up the disinfection process.

In her continuing experiments with the photocatalysts, Deepika also improved her formula for the concrete by replacing the zinc oxide with silver nitrate. The silver helps in three ways. First, it uses visible light as well as UV light in the photocatalytic process. Second, the silver absorbs electrons and helps increase the amount of ROS in the water. Third, silver kills bacteria, a fact humans have known since ancient times. For her Stockholm Junior Water Prize entry, Deepika documented studies showing that silver "binds to and alters bacterial nucleic acids, disrupts the cell membrane, and deactivates enzymes."

Her new, improved process involved running water through a pervious concrete disk mixed with photocatalysts. Then she let the water bottle sit in the sun with a photocatalytic disk inside. She found that this chemically infused concrete disk alone killed a significant amount of the bacteria. Water treated this way killed bacteria in just fifteen minutes. Deepika won the national Stockholm Junior Water Prize for 2014 for these innovations in the SODIS technique.

More Honors

Deepika continued to enter science fairs and win prizes. "I was really into competing," she says with a smile. She took first place in the environmental science division at the New Hampshire Science and Engineering Exposition in both 2013 and 2014. This qualified her for the ISEF in 2015, where she won a second-place grand award in environmental engineering. She also won a special prize for creativity.

In January 2015, *Forbes* magazine named Deepika as one of thirty people under the age of thirty who had made significant contributions that year. She knew the magazine had nominated her but had put it out of her mind. "Then one day, I was in class not paying attention and scrolling through Facebook, and on Facebook I saw a link by *Forbes*," she says. "So I clicked it . . . and I was shocked to see I was actually on the list."

That year she also won a Davidson scholarship, considered one of the most prestigious scholarships in the US. These scholarships provide funds to people eighteen and younger who have made significant achievements in the areas of STEM, humanities, and thinking "Outside the Box."

But the recognition didn't end there. "I was recently named one of the top twenty finalists [in] the international Google Science Fair," Deepika says. "We competed against each other and participated in fun activities. And we were able to meet Sergey Brin, one of the co-founders of Google. It was a great experience."

Deepika takes these prizes and awards in stride. "I think that as far as awards go [they] mainly give you a better platform to share [your] ideas," she says. "It's very humbling."

After graduating from high school, Deepika attended Harvard University. There, she joined several science-related organizations. For example, she joined the Scientista Foundation, which helps young women find role models and opportunities in STEM fields. Students for Climate Action, another of her affiliations, connects people interested in STEM to solve problems caused by climate change.

> People often think young people are apathetic or not capable of working on the massive issues the world is facing, but that has not been my experience.
>
> *—Deepika Kurup*

Deepika graduated from Harvard in 2019 with a degree in neuroscience. Reflecting on her Harvard experience, she says, "My classmates are also constantly inspiring me because they all have such passion to solve the problems they care about. People often think young people are apathetic or not capable of working on the massive issues the world is facing, but that has not been my experience."

Going Forward with Her Invention

Deepika sees many applications for her solar disinfection method to bring sustainable, affordable, and effective wastewater purification to developing nations. She has patents pending for her process and may form a nonprofit organization.

Deepika wants to see her invention deployed to help children like those she saw drinking contaminated water. Reflecting on the water crisis, she says, "When I was fourteen, I didn't realize that it would be so difficult to solve a global challenge. I just went for it."

To achieve this dream, she's met with organizations such as the Water Project, which provides water and sanitation in developing countries. Deepika would like to collaborate with organizations that are already working on these problems.

But Deepika still has to take many steps before she can deploy her invention and help solve this global challenge. She knows that she has to finish some more research and development before partnering with organizations like the Water Project. As Peter Chasse, president and founder of the Water Project, says, "She's got to test this technology [in the field]. How's it going to work in the real world?" Speaking from his experience with many clean water projects, Chasse continues, "Are people going to accept it? Are they going to use it? Will it last? And will it really disinfect the kinds of water we come into contact with?"

After talking to Chasse and others, Deepika comments, "These are things that I'm starting to learn." At first, her focus was totally on the science. But she sees how implementing even a great idea takes time. Working with the people in the community is key to success. "It is so different on the ground."

Deepika is convinced people can solve these global problems through partnerships between scientists and community action groups. "Reach out to as many people as you can, including the people who are directly affected by the issue you're trying to solve," she says. "It's important to start talking to people and working together early in the process."

Stanford Med

Continuing on toward her goal to become a neurologist, Deepika entered Stanford University School of Medicine in the fall of 2019. Following her white coat ceremony, the ritual that welcomes medical students into the profession, Deepika tweeted, "I feel so privileged to have the opportunity to learn how to care for others."

Because of the COVID-19 pandemic, most universities suspended their in-person classes and conducted courses through video conferencing software instead. Reflecting Deepika's sense of humor, her Twitter account lists her institution as "~~Stanford~~ Zoom School of Medicine."

Although Deepika studies neurology at Stanford, her concern for climate and water issues remains. Along with her passion for solving the global water crisis, she is very interested in promoting STEM education. "I attribute my quest for knowledge to my interest in STEM," she said. "I am a very curious person, and science provides the answers to many of my questions." These learnings lead us to revolutionary new discoveries that can solve global challenges.

Advice to Peers

How would Deepika advise other students interested in scientific research? "I would tell them that I really did get started just by googling things," Deepika says. She suggests learning all you can about what is already being done. She found that developing a hypothesis naturally led to experiments. As she learned, this experimenting can take a long time and breakthroughs require patience. "Research can be intimidating." But by following a scientific method and persisting, she succeeded.

As for tackling global challenges like the water crisis, she says, "Young people are the ones who will face these problems, and so young people will need to take action in order to solve them."

6

Cristian Arcega, Lorenzo Santillan, Oscar Vasquez, Luis Aranda, and Stinky, Their Champion Robot

Four teenagers huddled around a wooden frame in the Marine Science Lab at Carl Hayden Community High School. The object was a prototype of an underwater robot they were building to enter a NASA-sponsored contest. The contest presented a fictional story line in which a torpedoed submarine sank in the Caribbean Sea in 1942. The robots in the contest would need to explore a mock-up of the fictional sub in a swimming pool and assess damages.

The four teens—Cristian Arcega, Lorenzo Santillan, Oscar Vasquez, and Luis Aranda—animatedly discussed how to design and build a robot capable of completing a series of complex tasks. The robot had to locate an underwater signal beacon, recover two different pieces of "lost" equipment from the bottom of the pool, and measure both the length and depth of the "sunken sub." It also had to determine the temperature of a "cold spring" near the submarine and extract a precise amount of liquid from a sunken barrel. Any of these seven tasks seemed impossible, but this team succeeded beyond their dreams.

At first glance, the four teens appeared a most unlikely team, with different personalities, different interests and goals, and different skills. But they also had some commonalities. All of them were born in Mexico and living as undocumented immigrants in the US. They lived in West Phoenix, Arizona, an economically depressed part of town, and went to Carl Hayden Community High School. Carl Hayden was an underfunded high school, where almost all the students were Latinx and came from low-income families. All four were new to building robots, and they had little money to buy equipment. They would be competing with experienced, well-funded teams from around the country, including MIT, which had an $11,000 budget and an ExxonMobil corporate sponsorship. And whenever they traveled or even walked the streets of Phoenix, they risked arrest and deportation. All they had going for them was their ingenuity, their hard work, two teachers who believed in them, and one another.

As they worked together to solve their first engineering challenge, they tossed out ideas. The first challenge they worked on was measuring the length and depth of the mock sub at the bottom of the pool.

We could drop a string and then measure that length, Lorenzo suggested to the group.

Lorenzo

Lorenzo's mother brought him across the US-Mexico border through a tunnel when he was nine months old. He didn't fit in socially at school and others teased him unmercifully. So, he got into lots of fights, and the school assigned him to anger management sessions. His grades were low. Money was tight at home, and his family was on the verge of losing their house. Entering high school and looking for a group to belong to, Lorenzo joined a street gang. He soon pulled out when he saw his friends arrested for petty crimes. He didn't want to go to jail.

Lorenzo wanted to build things. His godfather operated an informal car repair service at home, where Lorenzo and his brother helped. From this experience, Lorenzo learned that when you don't have a full set of tools and all the supplies, you need creativity.

Cristian

"What if it doesn't reach the bottom?" countered Cristian.

Technology fascinated Cristian from an early age. At four years old he disassembled a radio, broke some wires, and then flipped the switch to see what would happen. The very satisfying loud pop and flash from the radio triggered the circuit breaker and the house went dark. *That was fun*, Cristian thought, even as his mother yelled at him.

At five years old, he announced to his family that he wanted to build robots. His parents, neither of whom had finished elementary school, didn't know where he got his obsession with robots. "He always wanted to get inside everything," his mother says.

In eighth grade, he and some friends built a rocket for a science project. Cristian had miscalculated the power of their engine. Setting it off at school, the rocket broke free of its guide wire and zoomed toward a group of kids playing soccer at the other end of the field. The rocket missed the group and caused no injuries. Cristian had to promise to be much more careful in the future. He was, however, already planning what he would build next. After that, he rarely miscalculated anything.

Oscar

"Yeah, that's a flaw," Lorenzo conceded. Thinking for a minute about the length problem, he suggested using a tape measure instead of a string. "We can tie a loop onto the end, hook it on to the submarine, and drive the robot backwards. The tape will just roll out."

"How will we read it?" asked Oscar, whose drive and indomitable spirit made him the natural leader of the team. He was mature, hardworking, strongly motivated, and deeply principled.

Oscar and his mother had wriggled through a hole in a chain-link fence on the US-Mexico border when he was twelve years old. They walked across the desert at night to join his father, who was already working in the US. His mother wanted to stay in Mexico, but she believed that Oscar would have better opportunities in the US. Oscar had already distinguished himself as a talented student.

In the US, Oscar again proved himself a strong student in school and even won a $200 prize in a middle school science fair. Arriving at Carl Hayden, he joined the Junior Reserve Officers' Training Corps (JROTC) and once again excelled. Pushing himself relentlessly and helping his teammates improve, Oscar inspired all around him. He was a motivator who led by example.

Promoted to the rank of executive officer, the highest rank in the JROTC battalion, Oscar dreamed of joining the US Army and defending his new country. In his senior year, when he inquired about enlisting, the recruiter shattered his hopes. He learned that without a green card, the army would not take him, no matter what he achieved in JROTC.

Lorenzo responded, "Aim a camera at it. We can read it off the video monitor."

"That could work," Oscar said encouragingly.

Luis nodded in agreement.

Luis

The last member of the team was Luis—big, quiet, Luis. In Mexico his family struggled to earn enough money. His father had been a construction worker but left for the US to make more money as a farmworker and send some home. His mother, who worked as a maid, realized that reuniting the family in the US would provide a better life and more opportunity for Luis. One day she and Luis, along with some other family members, took a bus to Nogales and crossed the border to the US through a hole in a chain-link fence.

To help his family out, Luis started washing dishes in a restaurant six hours a day after school. By high school, his father secured him a green card, and he was working as a short-order cook, and he found that he loved making food for people. School wasn't that interesting to him, but he wanted to respect the sacrifices his parents made for him.

A gentle giant, Luis had a 6-foot (1.8 m), 220-pound (100 kg) frame and a silent, steady stare, intimidating many people. But he didn't intimidate Oscar, who recruited him for the team.

Seeing the positive reception from his team in response to his idea, Lorenzo beamed with pride. So often outside this lab, people ignored or ridiculed Lorenzo's ideas. Here he had friends. And Oscar was building a new team.

One problem solved, many more to go.

The World outside the Lab

Just outside the walls of Carl Hayden Community High School was the stark world of West Phoenix, Arizona. A former Confederate soldier and his Mexican wife founded Phoenix in 1868. As the city developed, money for paved roads, sewers, and water lines went to the areas where white settlers lived. The areas where Mexican settlers lived got very little. During the economic boom of World War II, major companies including Goodyear and Alcoa brought many working-class white people to the community. The city originally built Carl Hayden for these families, and in 1965 most students were middle class and white.

In the late sixties and seventies, the white families began moving to East Phoenix and the city became increasingly segregated. The school fell into disrepair as the student demographics changed from nearly all white to nearly all Latinx. The Supreme Court ruled in 1974 that schools could not bus students out of their districts, yet a federal judge had ordered Phoenix to desegregate its schools.

Administrators tried to attract white students to Carl Hayden by turning it into a magnet school, specializing in marine science and computer programming. It didn't work. White families continued to move to East Phoenix, and immigrant families lived in an increasingly desolate West Phoenix. In 2005 the courts lifted the desegregation order. West Phoenix continued to decay economically, and the city neglected Carl Hayden.

The political climate in Arizona then was hostile to immigrants like these four teens. Politicians with national aspirations railed against immigrants, particularly those who entered the country illegally. Joe Arpaio, the Phoenix sheriff, vigorously pursued a policy of rounding up and deporting undocumented immigrants. He ordered his deputies to stop anyone who looked or sounded "Mexican" and demand to see their proof of citizenship. Authorities sometimes wrongly deported immigrants with documents, who then struggled to reenter the country. Police picked up one mother when they heard her simply speak in Spanish to her five-year-old on the street. Racism was pervasive and institutionalized.

Linking Up the Team

Fredi Lajvardi, an immigrant from Iran, created an extraordinary opportunity for students at Carl Hayden as the head of the marine science program. When he first came to Carl Hayden, he introduced the Science Seminar, a yearlong independent study project designed to allow students to dig deeply in an area of interest. This unstructured class allowed students to imagine, create, and pursue projects they wanted to explore.

One of his own teachers with her own science seminar had inspired Fredi when he was in high school. Obsessed with making hovercraft, Fredi built bigger and better hovercraft each year in her class. He began with a paper and balsa wood hovercraft powered by a small electric motor in his first year. By his senior year, his hovercraft was big enough for him to ride on and was powered by an old snowmobile engine able to hit speeds up to 25 miles (40 miles) per hour. He won first place every year at the regional science fair and attracted a lot of attention.

His parents, however, were unimpressed by what they considered a toy and a distraction from his schoolwork. They wanted him to become a doctor. Though Fredi really wanted to design and build machines, he dutifully started a premed program at Arizona State University. But he soon found himself drawn back to volunteering in his mentor's science seminar at his old high school. To Fredi, the boring lectures and memorization in the premed program were no match for the excitement of building things with students. Ann Justus, his mentor teacher, watching him work with her students, told Fredi that he was meant to be a teacher.

He finally listened to her, got his teaching credentials, and came to teach at Carl Hayden, where he took over the marine sciences program. There he met Allan Cameron, who headed the computer science program and was a kindred spirit. Both of them believed in inspiring students rather than lecturing to them. Both teachers chafed at traditional teaching methods and curriculum, focusing instead on sparking student excitement about learning with big challenges and firing their creativity. They had little interest in getting students to jump through tedious hoops of memorizing standard answers to standard questions for standardized tests. Both of them still remembered their own painfully boring experiences in school. Together, they coached the robotics team. They created the environment where Cristian, Oscar, Lorenzo, and Luis could thrive.

Meeting at the Lab

Sometimes the stars align to bring a great team together. Cristian took a marine science course because he heard it might give him a more intellectual challenge than his other classes, which he easily aced. And he heard that he could build robots.

Barred from a career in the army, Oscar signed up for marine science looking for a new challenge and a new team. When he met Fredi Lajvardi and heard about the robotics team, he realized he had found what he needed.

One day at school, as Lorenzo ambled through the halls, a poster recruiting for the robotics team grabbed his attention. When he turned up at the lab, Fredi showed him the impressive array of tools and told Lorenzo he could learn to use them all. Lorenzo was in.

Luis signed up for Fredi's Science Seminar in his senior year. Given his long work hours, sitting and reading at school often put Luis to sleep, so he wanted something active. Fredi offered Luis several options for his seminar project, including robotics. Luis chose robotics so he could build things.

Oscar recruited Luis for the team, partially for his strength. He could heft the heavy robot around as well as in and out of the water. But also, from years of work experience, Luis had proven himself steady and dependable. He was the kind of person Oscar wanted on the team. Together, Oscar told Luis, they would do great things. Never one to waste words, Luis simply said, "Okay."

These four teenagers—a brilliant student who nailed engineering challenges, an out-of-the-box thinker and creative builder who lacked direction and self-discipline, a high achiever and natural leader with dashed dreams of a military career, and a quiet hard worker who was always up to the task no matter how daunting—built a winning robot. Their immigrant status had blocked them all. They lived in a political climate hostile to their aspirations and an economic environment with little mobility for them. How did they do it?

These boys shared a desire to build things and to make a better life for themselves. They worked hard, they worked smart, and they worked together. And they had inspiring teachers to coach them.

Taking on Challenges

Fredi and Allan heard about the MATE ROV Competition, an underwater remotely operated vehicle (ROV) competition organized and run by the Marine Advanced Technology Education (MATE) Center and sponsored by NASA. In the summer of 2003, they attended a workshop for teachers interested in leading student teams entering the competition. They came back fired up since the contest combined Fredi and Allan's interests: marine science, robotics, and computer programming. The two teachers enjoyed giving students big challenges. And it wasn't as if they needed to win. The team could learn a lot and have fun. "Yeah, let's do it," they agreed.

Right away, the robotics team faced many challenges. Probably the most daunting were the technical challenges involved in building a functional robot that could maneuver underwater and perform tasks such as measuring the length of an object. These hurdles were considerable for all the teams, even experienced ones like MIT. Since none of the four on the Carl Hayden team had ever built a robot before, they faced a steep learning curve. Was it even possible for them to build a robot this complex and compete with more experienced clubs? They had so much to learn about robotics and about working together.

Beyond that, the team and their coaches had decided to enter the harder college division, rather than the definitely-hard-enough high school division. Why not aim high? They reasoned that if they were going to lose, they might as well learn all they could and lose to teams like MIT. Lorenzo laughed, "That can be our motto: 'Don't finish last!' "

Raising Funds

Along with getting started on the technical challenges, the Carl Hayden team needed to raise the money to build their robot. The school provided the tools, the teachers, and the facility, but the team needed to buy their own materials.

All their parents worked hard to provide the basics for their families, and they often had to make sacrifices to get by. For example, Cristian slept in a wooden box fastened to the side of his parents' single-wide trailer so his sister could have her own bedroom. Oscar and his family shared a small house with another family for a time. Lorenzo's family was in genuine danger of losing their home, having received an eviction notice. Most families in West Phoenix lived from payday to payday. None of their families could just write a check without a care. Lorenzo put it succinctly: "I don't know anybody with money."

Nonetheless, the team started asking family and friends for money—with little success. Then Luis stepped up and approached his manager at the restaurant where he worked nights as a short-order cook. He told his boss about the team and explained why they needed money. Respecting the hard, reliable work of Luis, the impressed restaurant manager gave him $100 for the team.

Inspired by Luis's success, Oscar asked the owner of the mattress factory where his father worked to contribute. Iris, the owner, knew Oscar since he worked at the factory during the summer. After hearing Oscar explain about their team and the competition, she wrote a check for $400, and another employee also chipped in

Legal Challenges

A group of nine honors students from Wilson Charter High School in Phoenix, Arizona, entered a science competition in Niagara Falls, New York. After winning a regional contest with their solar-powered boat, they earned the right to enter the international contest sponsored by the Institute of Electrical and Electronics Engineers. After the contest, their teacher took them to see Niagara Falls, one of the wonders of the world, and something they could not see in the deserts of Arizona. Everyone was having a great time, and the teacher wanted to take them to see the best view, which was on the Canadian side of the border.

She asked one of the border officials if the students could use their student IDs to cross the bridge, view the Falls, and come back. She reasoned that if her students weren't allowed to cross the border using their student IDs, then they would leave the visitors center and go back to their event.

Instead, the agent went to the visitor center and began quizzing the students about where they were from and

several hundred dollars. The team raised about $900 and learned how to ask for help.

This budget was not much compared to MIT's $11,000, but it was enough. They couldn't buy expensive electrical components or afford shiny machined metal body panels like many teams. They would use PVC pipe from Home Depot, spare parts they scrounged, and lots of the ingenuity Lorenzo learned while working on cars with his godfather.

And then, of course, there were the legal challenges. These boys were immigrants, and three of them were without documents. Whenever they traveled, they risked getting stopped and checked for documentation.

what their backgrounds were. He demanded to see proof of citizenship. Four of the students had been brought across the border illegally by their parents as children and had no documents.

The US border authorities immediately detained the four students, even though they had not tried to cross the border. Agents interrogated them for nine hours, and several agents addressed them with ethnic slurs. The students were told that they were being questioned because of their appearance. The border agents began the process to deport these students to Mexico. "Don't send your illegals to New York," one of the border officials told the school principal over the phone when she intervened.

They were allowed to remain with their families in the US during their legal appeal. Finally, three years later, a federal judge ruled that the overzealous border authorities had violated the rights of the students and dismissed the case.

Solving the Engineering Problems

For one of the challenges, the team knew that they had to measure the depth of the "sunken sub" as well as its length. They realized Lorenzo's clever solution for measuring the length wouldn't work for measuring the depth. The bottom of the pool would have nothing to hook the tape measure onto. As they discussed various approaches, Oscar asked, "What about a laser tape measure?" Since he had worked on construction projects, he knew about these devices that measured distances very accurately simply by pointing a laser beam, measuring the time it took for the laser to return after bouncing off an object, and displaying the distance on a screen. But would it work for objects underwater? Oscar didn't know.

"You guys should call somebody," Fredi advised. "The best way to figure something out is to call an expert."

After checking the internet, Oscar found Distagage, a company that offered high-quality laser tape measures, selling for up to $725 each. Should they even bother to call? "Just ask for advice," Fredi said.

Oscar called, reaching Greg DeTray, the owner of this one-person company, and explained their project and their problem. Talking to a high school student about the details of laser tape measures was a first for DeTray, and he was intrigued. When Oscar mentioned that they figured they might buy an inexpensive model from Home Depot, DeTray advised against it. Those laser devices only used the laser to pinpoint a spot and then used sound waves for the actual measurement. These devices often gave faulty readings.

Oscar asked if DeTray's laser tape measures worked for objects underwater. DeTray said he didn't know but would test it at home that night. The next day he called Oscar back and talked to the entire team over speakerphone. He had some bad news.

DeTray told them about using the device with an object in a big tub of water and taking some measurements. The results were clearly off. Each time he tested it, he got consistently wrong answers. The measurements were off by about 30 percent.

"The index of refraction!" Cristian blurted.

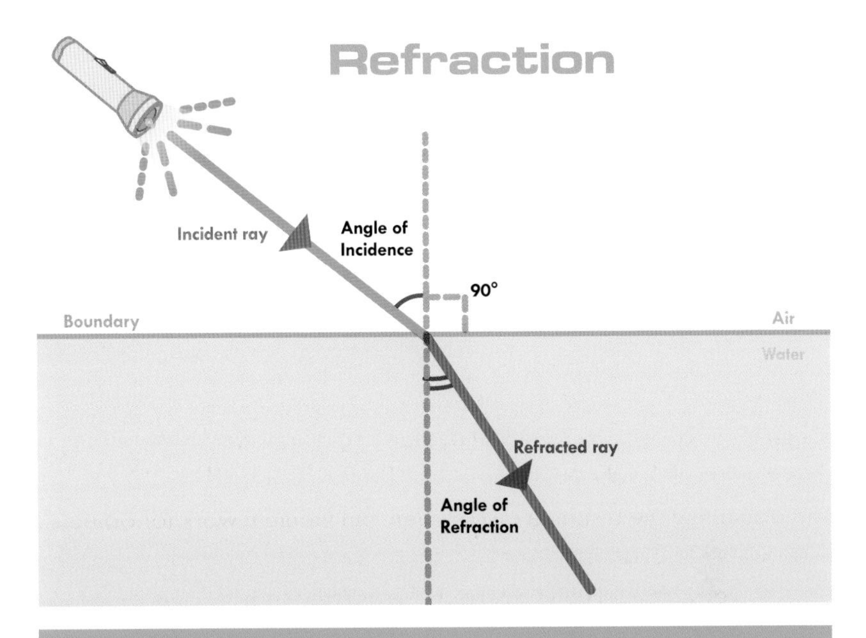

Light travels more slowly in water than it does in air. When a beam of light passing through air enters water, it slows down, causing a slight change in direction. This change is called refraction. Refraction makes objects in the water appear to be in a different place than they actually are. The team used an equation called Snell's Law to make the needed correction so their robot could find underwater objects.

Since water is denser than air, any light, even a laser, will travel slower in water than in air. The manufacturers designed the laser tape measure to work in air.

"So if we take 30 percent off the readings, it'll give us the right measurement," Oscar said.

"Exactly," Cristian replied.

Impressed by the team and how they solved a problem he hadn't figured out, DeTray offered to loan them one of his laser tape measures for the contest. "Thank you, sir," Oscar said. "We really appreciate it."

Another of the challenges was even stickier: measuring the temperature of a cold spring current flowing upward from the bottom of the "sea." The team wasn't sure how to do this, so Oscar searched the internet and found a company specializing in temperature-measuring devices. Oscar's call reached Frank Swankowski, a highly experienced temperature engineer.

Swankowski listened as Oscar explained the contest and their challenge. This was so different from his daily conversations with engineers and scientists. A few weeks earlier, another college team entering the contest had called and told Swankowski about the underwater ROV championship. They had just placed their order and hung up.

Oscar told Swankowski that his team was competing too, and that they wanted to learn as much as possible from experts like him. Liking this team's attitude, over the speakerphone, Swankowski gave them a thorough explanation of the science of thermometers and how to take the measurement they needed. Wanting to see this team win, Swankowski offered to donate the best device for their task.

After hanging up, Oscar faced his team. "You hear that?" he said triumphantly. "We got people believing in us, so now we got to believe in ourselves."

An Arm, a Hand

A few weeks into the ROV project, Fredi had arranged for the team to visit SeaBotix, a company in San Diego that built underwater robots. Donald Rodocker, the company president, had built and designed underwater robots while he was in the navy and was a legend among ROV builders.

Rodocker took them on a tour of his facility and the lab where the latest creations were taking shape. The boys were in awe of the robots and the tools. They also carefully examined the racks of grippers and mechanical arms, thinking about the two challenges that required their ROV to pick up objects from the bottom of the "sea."

At the end of the tour, encouraged by their experiences with DeTray and Swankowski, Oscar asked Rodocker if they could borrow one of the prototype arms. *Sure,* he agreed. And just like that, the boys had an arm that would be perfect for grabbing the captain's bell and retrieving the towfish, the sonar beacon. Things were falling into place.

But on the way back to school, they all had a sobering experience. Just after crossing into Arizona from California, they came to an impromptu Immigration and Customs Enforcement (ICE) checkpoint. Thinking of the experience of the Wilson High students (see sidebar on pages 140–141), everyone was frightened. Would ICE arrest them and deport them? Their families, their homes, and their lives were in the United States.

Fredi told them to give him their school IDs and to keep quiet. He would do all the talking. As they rolled up to the checkpoint, the officer looked over the team and examined their IDs. Fredi told him they were on a school trip. The officer took another look around and waved them through. They were okay this time, but the experience sent a chill through them all. This ever-present danger would always be on their minds as they traveled.

Designing and Problem-Solving for Waterproofing

Back at school, the team set about building their robot. Many of the college teams would have a machined metal body to house the electronics and other devices. Carl Hayden did not have a machine shop and, even if it did, none of them knew how to work such machines.

Cristian had done some research and suggested that they use glass syntactic foam, made of glass microspheres embedded in an epoxy resin. James Cameron had used that material for his film *The Abyss*, a science-fiction thriller about an attempt to recover a sunken sub in a deep underwater trench. The syntactic foam sounded good to everyone until Cristian mentioned that it would cost a couple of thousand dollars for the amount they would need. In an ideal world with a big budget, that would be the way to go.

Once again, they would need to be creative. They needed a cheap, sturdy, waterproof material. Metal was out since it was expensive, and they couldn't shape it. Wood would be great as it was sturdy, relatively cheap, and easy to work with, but making it waterproof would be difficult.

What about making a frame out of PVC pipe? It was cheap, sturdy, waterproof, and a material they were familiar with. Lorenzo pointed out that they could run the wires inside the pipes to stay dry, and since the pipes were hollow, the frame would float.

Cristian grabbed some paper and began doing some calculations. A PVC frame and all the air inside would make the robot too buoyant. We'll need to add weight, he told the team. What if we put the battery on the frame for added weight?

The idea of putting the battery underwater was a bold suggestion. Just one little leak in the protective cover and the power supply could short out and disable the robot.

But this solution had several design advantages. They wouldn't need extra weight, which would take up valuable space on their robot and make it harder to pilot. With the battery on board, they could have a thinner tether to the surface, making the ROV more maneuverable. And the shorter tether would reduce the amount of power lost in transmission. Their battery would last longer.

Should they take the risk? Oscar pointed out that the other teams wouldn't do this, since a leak was a contest-ending danger. "If we do the same thing as everyone else, we'll finish last," Cristian said, "because they've done it before."

Luis spoke up, "He's got a point."

That decided it. The battery went on board.

The Mystery Liquid Problem

One of the toughest challenges in the contest was sampling 17 ounces (500 mL) of a "mystery liquid." The story line was that when the sub sank, several barrels fell to the seafloor and were leaking an unknown and possibly hazardous liquid into the ocean.

Lorenzo took on that task, since Cristian and Oscar thought it might be too hard. They wanted to concentrate on the easier tasks that would earn them points.

The college teams they were competing against devised some complicated solutions that included using electronically controlled three-way valves, multiple pumps, and syringes to pump the mystery liquid into rigid containers. Lorenzo decided to use a balloon. It was cheap, flexible, wouldn't add buoyancy, and could expand as needed.

The team found a simple pump at Home Depot designed to remove water from flooded cellars. It was small, cheap, and powerful. Could Lorenzo make it work?

He hooked up each end of the pump to some PVC pipe and then glued some copper tubing at the ends. The copper tubing on the intake side was small enough to fit through the opening in the barrel. Lorenzo bent it so they could maneuver it into place. He fitted a balloon to the end of some more copper tubing on the other end of the pipe where the liquid would emerge.

When he tested it, the pump and balloon worked perfectly. By timing how long to run the pump, he could measure out 17 ounces (500 mL) of liquid. But as the balloon filled, it got so heavy that it slipped off the end of the copper tubing, spilling the liquid.

After trying several ways to fix the balloon to the pipe, none of which worked, he was getting frustrated. All he had to show for his efforts were wet shoes. Then he tackled the problem a different way. *What if I make a container to hold the balloon in place as it fills up?*

He pulled a large soda bottle from the trash, cut it in half, and fixed it beneath the balloon. It worked fine until it was nearly full. Then the swollen balloon stretched over the edge of the soda bottle and burst. The next day, he brought in a 1-gallon (3.8 L) plastic milk container, cut off the top, and put it beneath the balloon. When he turned on the pump, the balloon filled, and the milk container held it in place. Simple. Cheap. Effective.

Propulsion Help

How would they power their robot and move it around in the pool? Once again, they could not afford expensive electronic motors and did not have the time or skills to build them from scratch. Fredi suggested using the electric motors used for trolling—fishing from a slow-moving, silent boat. These trolling motors were small and used little power and so they would not drain the battery so fast.

The team worked out that they would need five motors to move the ROV in all directions, including up and down. Luis reminded everyone they also needed to tilt the ROV so they could use their mechanical arm to pick up objects, such as the captain's bell and the towfish.

After checking the internet, Oscar got on the phone and reached a manufacturer's representative, Kevin Luebke at Mercury Marine. Luebke was used to talking to people who wanted to fish, so Oscar needed to explain what they were doing and why they needed motors. Oscar's story fascinated Luebke, and he wanted to help. He sold the team the motors they needed at a hefty discount. Then the robot had a propulsion system.

Setting up a tub of water and a piece of wood, they worked out where to place the motors for maximum maneuverability. Experimenting by pushing the wood around in the tub, they taught themselves the concept of torque by direct experience.

To hold the ROV's control system, they needed a waterproof container. They found a waterproof briefcase on sale at a local electronics store. Back at their lab, they drilled the holes they needed for wires, sealed the spaces, and tested the case in one of the big lab sinks. No leaks.

Then they began assembling the PVC frame and sketched out a plan. Cutting the PVC pipe proved very difficult for Cristian, Lorenzo, and even Oscar. But Luis wielded the pipe cutter with ease. "It's like butter," he said.

With the pipe cut, they assembled the frame. They glued nothing until they had worked everything out exactly as it needed to be because the glue was permanent. Their calculations proved accurate, and they then had a slightly awkward-looking robot frame sitting in their lab.

First, FIRST

Fredi and Allan wanted to give the ROV team some more practice in building a robot and some practice in competition. They enrolled the team in one of Dean Kamen's regional FIRST Robotics Competitions. All the boys learned how to solder and wire. A nice bonus was that each team in the FIRST competition got a carton of supplies, including a robot controller box. They would later repurpose this controller for their ROV.

As they thought about how they would build their FIRST robot, they went over the challenge. The robots would get points for picking up Styrofoam balls strewn on a big floor and then shooting them through basketball hoops. The more baskets, the more points. They could also get fifty points if their robot was the first to get to a pull-up bar on one side of the court and hoist itself to do a "pull-up."

The boys thought about the complexities of building and maneuvering a basketball-playing robot. All the robots on the floor would jostle and crash into one another, trying to get balls, make baskets, and block one another. This would be their first complete robot and their first competition. They would compete with many affluent schools with well-established, well-funded teams. And they didn't want to finish last.

As he often did, Lorenzo came up with a creative solution. What if, he said, we don't play the game? What if we just go straight to the pull-up bar? At first, it sounded crazy to Oscar, Luis, and Cristian. But as they thought about it, it began to make sense. The pull-up was worth a lot of points, as much as scoring ten baskets. By going directly there, they could be first and have no competition for the bar. They could focus all their energy on doing that one thing and not worry about building arms that could pick up round balls and then maneuver well enough to "shoot" accurately. Once they had the bar, no one else could do the pull-up.

So the boys from Carl Hayden went to work on building a simple robot optimized for the pull-up segment of the competition. This strategy worked. At the competition, while the other teams looked on in shock and surprise, the Carl Hayden robot just lumbered over to the pull-up bar and hoisted itself up. *Can they do that? Is that legal?* They could. It was.

They didn't win, but they were far from last. They came out in the middle of the pack of about seventy teams. This was enough to win them a trip to Atlanta for the national competition. This was a great honor, an impressive debut for the team, and lots of fun. But they also worried that during their travels, immigration authorities might discover them at some checkpoint. The case of the Wilson High students was still in the courts.

At the national championship in Atlanta, the Carl Hayden team followed their strategy and gained more competition experience. At the nationals, they also finished about midway among all the competitors, a very respectable showing for a new team. They felt good about themselves and their abilities. And they made it to the competition and back home uneventfully.

The Build of Stinky

Back in the Marine Science Lab, the team set about building their ROV. Since they only had one chance to get the gluing right, Oscar had the team practice assembling the robot over and over. Once applied to a pipe, the glue would dry quickly, and they needed to fit each pipe segment in the right place in seconds. They got their assembly time down from an hour to around twenty minutes with no mistakes. Then it was time to glue.

The glue for the PVC pipe had strong fumes. The fumes filled their workroom and stung their eyes and lungs. Their practice paid off, and they took turns gluing so that no one spent too long in the small room. The finished frame looked good.

But as they began to put the waterproof case in place, they saw that the legs did not exactly line up with the holes they had cut. The legs would not fit through. What to do? Start over? That would mean more time and money that they didn't have. *Maybe we could cut off one section,* Oscar suggested.

As he and Cristian considered how best to do that, Lorenzo looked closely at the case, the holes, and the legs. The holes were not that far off. "What if we just bend the pipe?" he said.

How can we do that? Oscar asked.

"The electric heat gun," Lorenzo replied.

An electric heat gun, which looks like a blow dryer, is used to dry paint and gets very hot. They tried the heat gun, and the PVC pipe softened enough that they could bend it the slight amount they needed. Then the legs fit just right. Another problem solved.

Finally, they needed a name for their ROV. Thinking of the powerful smell of the glue, Oscar said, "Why don't we call it Stinky?"

They painted Stinky bright blue and bright red, partly for decoration, as Lorenzo thought the plain white PVC looked drab. With the heavy battery and the electronics loaded, Stinky's frame could break if not handled carefully. Lorenzo color-coded the most fragile sections with red paint so they could avoid grabbing those parts when trying to lift Stinky.

After the assembly and the paint job, Stinky was a bit ugly and garish compared to the polished and trimmed robots of the other teams. But the competition would be the real test of all the robots.

Pool Practice

The team needed to practice using and maneuvering Stinky. Carl Hayden didn't have a pool, so they arranged with the friendly manager of a nearby diving school to use their pool during off-hours.

At first, as they got used to the controls, they bumped into the pool walls and worried about damaging Stinky. As they practiced, each of them learned their parts, and they began functioning as a team. Cristian and Oscar drove the motors that moved Stinky around the pool. Lorenzo managed the sensors and operated the robotic arm, the camera that was their eyes underwater, and the pump for sampling the mystery liquid. Luis managed the tether of wires and cables that connected Stinky to the poolside controls. By their second practice session, they were getting good at using the tape measure to assess the depth and using the claw to pick up PVC pipe off the bottom of the pool.

One task they couldn't seem to get right was the liquid sampling. The pump worked well, and Lorenzo could suction 17 ounces (500 mL) of liquid in just twenty seconds. The problem was maneuvering Stinky into place so that the copper intake tube would connect with the half-inch (1.3 cm) opening in the barrel. They couldn't manage the fine movements precisely enough to insert the tube. But with everything else working well, they plunged ahead.

Finally, on the appointed day, the team met at the school and loaded their vehicles with Stinky and all their equipment. They drove out of the school parking lot and headed west to Santa Barbara, California, and the MATE ROV Competition. Along the way on their seven-hour drive, they quizzed one another on the information they needed to know for the technical judging. This technical review would count for almost half the points in the competition. Each of them needed to explain their robot and their design decisions to the panel of engineers and robot experts from the navy and NASA. Questions might include, What's the index of refraction? What's a pulse-width modulation (PWM) cable? What transmitter frequencies are you using? Why those frequencies? The team knew the answers.

The Contest

Arriving in Santa Barbara, everyone was a bit stiff and tired from the long road trip but excited to be there. They checked into their assigned dorm room and unpacked Stinky for testing. At first, Stinky responded quirkily to the controls and then didn't respond at all. As they continued troubleshooting, Stinky started working again. Weird. Relieved, but worried by this glitch, the tired team turned in for the night.

The Leak

Early the next morning, they brought Stinky down to the practice pool for an underwater test drive. They would rehearse working the controls, get used to the campus and the other contestants, and relax. Only it didn't work out that way.

Shortly after Luis carefully lowered Stinky into the pool, the robot began responding erratically. When Cristian and Oscar tried to go straight, Stinky turned left. Cristian, sensing something very wrong, yelled to Luis to get Stinky out of the pool.

On the deck they opened Stinky up and began searching for the problem. Was it a loose connection? Was it a bug in the coding? Then Lorenzo spotted the problem, pointing to a small patch of water on the bottom of the case. A leak!

On the road trip, one seal in the case housing jarred loose, leaving a gap. They needed to completely rewire Stinky that night. And they needed to fix the leak.

What were they going to do? Oscar wanted to fix it right then, but Fredi pointed out that the team had their technical review in a couple of hours. They needed to focus on this review since it was worth so many points. Once realizing they had all night to fix their robot, they began preparing for their technical review.

Pitch Practice

Allan had an inspired idea to break tension and get in some practice. None of the boys were used to making public presentations. Allan suggested they go around the campus and explain their robot to anyone who would listen.

At first, all of them, including Oscar, were afraid to approach anyone. Would the wealthy white people here think they were panhandling? Would they just walk away? Lorenzo was the first of them to master his nerves and approached one man who looked professorial. "Hi, we're high school students from Phoenix, and we're here to compete in an underwater robotics contest. Do you want to hear about it?"

The man genially agreed. Then Oscar stepped forward with their three-ring binder, and the boys began their spiel. They spoke to several more people about their cool robot and drew confidence from how interested and impressed these people were. The boys felt ready.

The Technical Review

At the appointed hour, the boys filed into the room with the technical judges. Tom Swean, a grizzled veteran of the Office of Naval Research who developed underwater robots for the Navy SEALS, headed the three-person panel.

How will you measure the length? he asked. Cristian told him how they viewed the readout of the laser tape measure and then manually corrected for the index of refraction. The panel continued to pepper them with questions about their design and the technical specifications of their robot. Cristian excelled at this type of quizzing, but Oscar and Lorenzo also confidently fielded these questions. Lorenzo proudly pointed out some of their innovations.

You seem comfortable with the metric system, Swean noted after asking Oscar about signal interference.

"I grew up in Mexico, sir," Oscar said.

Where's your PowerPoint? Swean then asked.

"PowerPoint is a distraction. People use it when they don't know what to say," asserted Cristian.

"And you know what to say?" pursued Swean.

"Yes, sir," declared Cristian.

Meanwhile, Lisa Spence, who worked with very sophisticated robots as a part of NASA, noticed that Luis hadn't said much.

You used PWM cable in your robot. Why? she asked, looking directly at Luis.

During a long second, Cristian held himself back from jumping in. Then Luis calmly responded, "It stands for 'pulse-width modulation' and it is used for controlling high-current devices digitally." The team said a silent thanks—they knew Luis had nailed that question.

Meanwhile, Fredi and Allan paced and fidgeted outside the room, waiting for the boys. They knew that most teams spent around forty-five minutes in their technical review. When the boys came out all smiles after about twenty-five minutes, their teachers looked surprised. Had they totally blown their tech review? *How did it go?* they asked.

Great! the boys replied, full of confidence. Still not convinced, Fredi and Allan decided to let it go. It would be what it would be. And everyone knew they still needed to fix the leak and rewire Stinky.

The Leak Fix

Before tackling the leak and the rewire, the team and their teachers headed over to a restaurant for dinner. On the way there, Oscar led them in brainstorming workable solutions for the leaky case. They knew it was too late to buy a new case and fit it to their robot. They needed something to soak up water from any new leaks.

Cristian pointed out that their fix had to be small enough to fit in around the electronics in the case and it had to be superabsorbent.

Superabsorbent? thought Lorenzo, recalling TV commercials he had heard. *What about a tampon?* he offered. The other boys laughed nervously and blew off the idea.

Sounds like a tampon would work perfectly, said Fredi.

After a full meal of steak and shrimp, the team drove to a Ralph's grocery store near campus to buy some tampons. Since it was his idea, the team made Lorenzo go into the drugstore to buy the tampons. *How will I know what to get?* he asked. *Ask someone*, said Oscar. Lorenzo walked into the store with trepidation. After wandering a bit, he approached a woman shopping for shampoo. Taken aback at first, she then smiled as she listened to his story. She helped him find the right aisle and showed him the best brand for their purpose. Relieved, he headed to the counter to make his purchase as the woman called after him, "I hope you win."

After some quick tests back in their dorm room, they knew the tampons would work. Relieved and with full stomachs, the boys were tired and ready for bed. But Stinky still needed to be fully operational before dawn.

All-Nighter to Rewire

After the team assessed the damage, they knew they needed to completely rewire Stinky. Someone needed to resolder and check all the connections. This was a big and delicate job. Everyone was so tired.

"I'll stay up and do it," Oscar volunteered.

I'll help you, Lorenzo said quickly, determined to show he was reliable and committed. This offer surprised Oscar. He had seen Lorenzo as a bit of a goof-off, always with a silly joke or an off-the-wall idea. But he had to admit that many of Lorenzo's ideas, which sounded preposterous at first, worked out.

The two set to work while the others slept. It was tedious, and they needed to work carefully. Accidentally touching the soldering iron in the wrong place would mean they would need to start all over. Around 2 a.m., Oscar and Lorenzo took a quick break. They were dead tired, but they knew they needed to finish. Any mistake at this point and there would not be enough time to rewire everything before the contest. They forged on and finished the job. Exhausted, they powered up the robot and checked their handiwork. Stinky was operational again.

The Main Event

Morning came quickly, and the team headed down to the university's outdoor pool to get ready for the main underwater event. All the other teams were also setting up. It was a brilliantly sunny Southern California day. Loud Hawaiian music blared over the thrum of activity on the deck. Allan dubbed one beautifully polished metal and shiny fiberglass robot "underwater jewelry."

Monterey Peninsula College was the first contestant of the day. Their team had three vehicles, each with a specialty. One craft simply floated on the surface with an underwater camera and could guide the action below. The other two vehicles were ROVs. They optimized the large bulky one for lifting heavy objects and the smaller, streamlined bot could more easily maneuver in tight spaces. Even with this sophisticated strategy and a team of fifteen people, Monterey only scored 30 points out of the possible 110 points during their thirty-minute allotment. These challenges were no cakewalk.

The other colleges that followed also had some very cool design features. Cape Fear Community College added an air chamber to their ROV that a scuba tank on the deck could fill. This gave their ROV

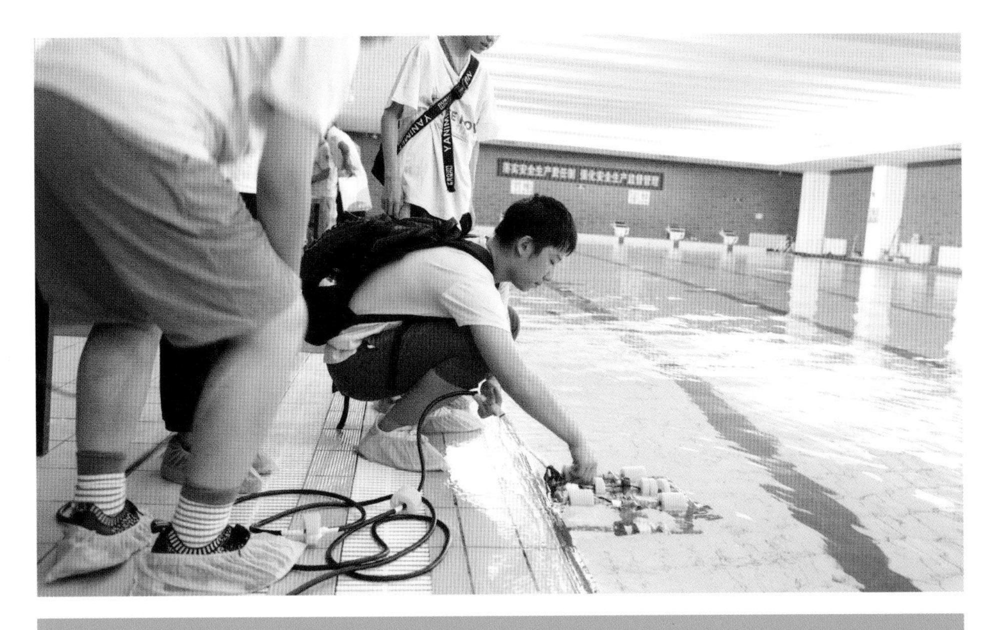

Each year, the MATE ROV competition comes up with new challenges. In 2022, it asked competitors to highlight sustainability and green energy because of the ocean's vital role in the planet's climate and ecosystems. Some of the tasks included planting and harvesting a seagrass bed, locating and measuring the length of a wrecked ship, and detecting the presence of marine mammals.

impressive ability to manipulate its buoyancy by adding air or letting it out. This Florida team did better than Monterey, garnering 40 points. This underwater challenge course was proving tough to beat.

Throughout the morning, the other teams also tackled the tricky challenges. No one so far was scoring very many points. Some robots just plunged to the bottom and didn't move. Others leaked and shorted out.

Then it was time for MIT. This team, expected by many to win, boasted a very talented group of programmers, mechanical engineers, and electrical engineers. With their $11,000 in corporate funding, they could afford the best materials, such as syntactic glass resin, a strong and lightweight material. They had worked hard and tried innovative approaches to the technical challenges, often using exotic new equipment they bought or built.

In the pool, their robot cruised efficiently around, picking up point after point. They did have trouble with the mystery liquid challenge. The nozzle of their complicated dual-container apparatus didn't fit into the opening on the barrel of liquid. Although they had to scratch that task, they still scored 48 points.

The Carl Hayden team watched from the deck, noting the difficulty of the tasks. Then it was their turn. At first, they had difficulty with maneuvering. Right away they spotted the towfish—a sonar beacon and one of the items to pick up—and headed for it. Overshooting, they had difficulty getting back to the spot. They moved on and tried the measuring tasks.

They looped the tape measure on one end of the "sunken sub" and moved back to a perfect spot to get the measurement. When they turned on the camera to take the reading, the bright sun created too much of a glare to read the tape. The same thing happened when they maneuvered in place to measure the depth. In their practice sessions, their laser measurement system had worked well because they were either operating in an indoor pool or because the day had been overcast.

Since each challenge assigned points for completing different parts of the task, the team picked up 5 points each for the two measuring tasks and 5 points for locating the towfish.

They went on to the mystery liquid challenge. In practices, they had great difficulty getting the nozzle in the hole on top of the barrel. But this time, Oscar and Cristian maneuvered Stinky into position and the nozzle slid right into the opening on the barrel. Lorenzo quickly started the pump and counted off the seconds as the balloon filled with red mystery liquid. They brought up their balloon and ran to measure it. They had 17 ounces (500 mL) of mystery liquid, with only a bit of dilution. For this challenge, they earned 12 of the possible 15 points.

"Can we make a little noise?" Cristian asked a judge and received an affirmative response. After a quick dance on deck, they zipped back to their stations to get as many more points as they could in their remaining time. At the last minute they spotted the captain's bell and grabbed another 5 points.

At the end of the underwater event, they had scored 32 points and were now in third place, behind MIT and Cape Fear. They couldn't believe it. Not only had they not come in last, but they had scored more than most of the college teams. They were thrilled.

"Congratulations," Allan told the stunned team. "You officially don't suck."

The Award Ceremony

That night at the award ceremony, the boys still basked in satisfaction with their performance. Their coaches, also pleasantly surprised, told them they would probably get fourth or fifth place, about the middle of the pack.

After the dinner, the MC announced the awards. *The judges created this first award spontaneously,* he said, *to honor a team for special achievement. This winner,* he continued, *was Carl Hayden Community High School team.* The boys happily came onstage to receive their award, but they couldn't help but feel that the judges intended it as a consolation prize. OK, so they probably hadn't finished third, they thought. The ceremony continued with some other awards.

The Design Award, one of the major prizes, to the surprise of many, including the boys from Carl Haden, went to Stinky. How could that be? The other robots were so cool looking and had such sophisticated equipment. As Lisa Spence later explained to them, a true test of engineering is to prefer a simple solution over a complicated one that might fail for many reasons in real-life situations. At NASA, they had learned that lesson the hard way. Spence saw that many of the boys' design solutions were conceptually similar to those used in the space program.

Then, before they could sit down, they were announced as the winners of the Technical Writing Award. Again, they were floored. Cristian had worked hard on the paper. He was brimming with confidence, knowing it was good, but to beat MIT in technical writing?

A kid from West Phoenix? Stunned and overjoyed, they had far exceeded their expectations.

Finally, it was time to announce the overall winners of the contest. Bruce Merrill, the master of ceremony, announced the third-place winner: Cape Fear. Everyone clapped and cheered, while the boys' hopes for third place sank. Monterey Peninsula must have edged them out in the technical review, they whispered.

Then Merrill announced that MIT had placed second. "MIT got second?" Cristian blurted out, expressing the surprise of many in the room.

"So who won first place?" Lorenzo asked the table.

Merrill began a drumbeat on the podium, and the audience joined in. Then he stopped and the room fell silent. *The grand winner is,* he shouted, *Carl Hayden!* Dumbfounded, the team and their teachers made their way to the stage amid loud cheering. They had done it.

The students from MIT stood up clapping, starting a standing ovation for the Carl Hayden team's accomplishment. This ragtag high school robotics team from the desert had just won the collegiate division of the 2004 MATE ROV championship. The atmosphere in the room was electric.

Overwhelmed, the team left the hall as quickly as they could after accepting their awards and their congratulations. They had overcome huge odds and exceeded their wildest dreams. They went down to the beach and shouted to the ocean.

"We won!" Oscar yelled.

Luis bellowed, "Yessss!"

"We beat MIT!" Cristian exulted.

They yelled and whooped until they were hoarse. Then they headed off to bed.

But on the beach, Oscar, along with his joy, was feeling fear. His eighteenth birthday was in a few weeks, and once he passed that milestone, the penalty for illegally being in the US increased. If he was caught then and deported, he could be banned from reentering the US for ten years. What should he do? The US was his home.

Coming Back to Life in Phoenix

After the rousing acclaim at the MATE ROV awards ceremony and their own private celebration of their victory on the beach, the team returned to Carl Hayden. No marching band welcomed them, and no school assembly honored their achievement. Almost no one noticed. Cristian and Lorenzo resumed their studies and merged back into their normal life at school.

Having graduated from Carl Hayden that spring, both Oscar and Luis went to work. Oscar worked as a day laborer on construction projects for about seven dollars an hour. As he worked, he dreamed about going to college to study mechanical engineering, serving in the military, and having a career as an engineer. As he carried sheetrock, he studied the plumbing and electrical wiring around him to keep learning. Someday.

Luis worked two jobs: during the day he was a filing clerk in the Social Security Administration office, and at night he continued as a short-order cook at the same restaurant. He had briefly experienced a world of challenge, excitement, and opportunity with the robotics team. Those bright prospects seemed to dim as time passed.

Then something happened. Freelance writer Joshua Davis heard about the team and investigated their accomplishment. He wrote the story of the improbable underwater robotics team from the desert in *Wired* magazine. This article generated a tremendous outpouring of interest, congratulations, and support from around the country. A scholarship fund for the team raised about $120,000.

Then ABC's *Nightline* wanted to interview them for a nationwide broadcast, highlighting both their achievement and their immigrant status. Despite the significant personal risks, the team decided to speak out. "This is a Rosa Parks moment," Oscar said. "It's about more than us now."

Oscar and Cristian used their shares of the fund to attend Arizona State University (ASU). Oscar immersed himself in his studies of mechanical engineering and founded ASU's first robotics team. Along with his studies and his robotics activities, Oscar worked in the community to encourage younger students to stay in school and to form their own robotics teams.

At Oscar's ASU graduation, President Obama spoke of hope in troubled times to the largest graduation audience ever. Oscar was one of three seniors specially commended for leadership and contribution to the community. Following graduation, married and with children on the way, Oscar thought about his future. Regardless of his accomplishments, his values, and his character, his legal status as an undocumented immigrant limited his career options. He was also in constant jeopardy of being deported.

Cristian also faced hard choices. Many people, including Fredi and Allan, encouraged Cristian to go to MIT, which was a natural fit for him. But his family, mindful of his legal status, feared for him living so far away. Agreeing to their wishes, Cristian signed up at ASU, but soon found listening to lectures about concepts he already understood painfully boring.

Then, in 2006, the Arizona legislature passed Proposition 300. At this time in Arizona and around the country, some politicians fanned the flames of resentment against undocumented immigrants. They claimed that immigrants from Mexico were taking jobs, receiving welfare benefits, and getting college subsidies that only American citizens should receive.

Like the claims of demagogues throughout history, many of these arguments were inaccurate, distorted, and oversimplified. None of these impassioned speeches mentioned the talents and skills of immigrants from Mexico. Studies have also documented the net benefits to the United States of immigrants—economically, culturally, and socially. Immigrants often start new businesses and create inventions as well as reinvigorate communities.

Proposition 300 legislated that undocumented immigrants were no longer eligible for in-state tuition at state colleges and universities, such as ASU. This made it impossible for many students to attend. After the law passed, Cristian's tuition quadrupled, increasing from about $2,000 per semester to about $8,000. He could not afford this and left school. He took a job as a store clerk at Home Depot, attended community college classes, and set up a workshop at home. In the evenings, he worked on his own inventions, using discounted or scavenged parts. He began saving money to return to school, hoping to earn a degree in engineering eventually.

Cristian also was in danger of being deported. Sheriff Joe Arpaio ramped up his plan to deport as many immigrants as he could. Arpaio formed citizen posses to report people suspected of being undocumented immigrants. He ordered his deputies to strictly enforce traffic laws and then ask for proof of citizenship from any Latinx people they stopped. No one asked white drivers for any citizenship papers.

Different Futures

Following his graduation from Carl Hayden, Lorenzo pursued his best option and made his interest in food his career. Both he and Luis used their scholarship money to attend cooking school. Then the two friends set up a catering business, working parties and events on weekends. During the week, Lorenzo worked as a line cook at a high-end restaurant. Meanwhile, Luis found work as a night-shift janitor at the federal courthouse and worked his way to the position of janitorial supervisor.

On the other coast, the members of the 2004 MATE ROV team from MIT had different experiences. Their tech wizard graduated and got a job at a robotics firm. The team captain got his PhD and went on to build underwater robots for the research facility at the Monterey Bay Aquarium in California. Another team member works for NASA and went to Antarctica for a project.

All the members of this smart and talented team deserved their excellent positions, wonderful careers, and secure lives. But it is hard to ignore the very different life outcomes of the members of the two teams, based in great part on the very different starting points and privileges they got in life.

Becoming a Citizen the Hard Way

Oscar had funded his increased tuition at ASU through working jobs and from funds provided by groups at the university who recognized his accomplishments. Luis also donated some money from his share of the scholarship fund since his schooling was less expensive.

After graduating with a bachelor's degree in mechanical engineering, Oscar realized he was at a crossroads. His immigrant status made his life precarious. It prevented him from fully taking part in the community and from pursuing his career goals. He decided to become a US citizen.

His first step was to deport himself to Mexico—the country where he was born but was no longer his home. He crossed the border at El Paso, Texas, into Juarez, Mexico. At the consulate in Juarez, he applied for residency in the US. Since he truthfully admitted to having lived in the US illegally, they denied his application for residency. Oscar received an official letter saying that he could not reenter the US for ten years.

In Mexico, cut off from his wife and daughter back in Phoenix, Oscar first supported himself by picking beans in the fields. This backbreaking work in the scorching sun paid about four dollars a day. After a long search and help from a relative, he eventually found a job at a car parts factory supervising assembly line workers.

Senator Richard Durbin of Illinois, a sponsor of the Development, Relief, and Education for Alien Minors (DREAM) Act, learned of Oscar's situation from immigration activists and letters written by Fredi, Allan, and his many supporters. Durbin interceded with the immigration officials to review Oscar's case, considering his many accomplishments and engineering skills. As a result, they granted Oscar a permanent residency status in the US, and Oscar came home to his wife, Karla, and their baby girl, Samantha. Although offered a job working on robots in Phoenix, Oscar followed his long-held goal and enlisted in the US Army. Based on his engineering degree, they offered him a spot in Officer Candidate School. Oscar turned it down since he wanted to earn his leadership position.

What Are the DREAM Act and DACA?

Some immigrants entering the US without authorization bring their young children with them. Officials estimate there are about eight hundred thousand such undocumented young people living in the US. The DREAM Act provides them with a path to citizenship. Like Lorenzo, Oscar, Luis, and Cristian, these undocumented youths have lived most of their lives in the US and consider it their home. One young man, Jose Antonio Vargas, learned that he was undocumented when he applied for a driver's license. He was shocked when the clerk told him the green card given to him by relatives was a forgery. She denied his application and advised him to leave immediately. Without citizenship, undocumented people face challenges getting jobs, attending college, and receiving community services. They cannot vote, despite misinformation to the contrary. They are constantly at risk of discovery by immigration authorities, arrest, and deportation.

Undocumented immigrants cannot draw benefits from the Social Security system or Medicare. But they contribute an estimated $13 billion to Social Security and $3 billion to Medicare through payroll taxes, according to NPR's *Marketplace* podcast. The Institute on Taxation and Economic Policy estimates that undocumented residents of the US pay about $12 billion each year through state and local taxes. Many studies, from sources such as the Congressional Budget Office, the Cato Institute, the Wharton School of Business, and the Brookings Institution, show that immigration reform will grow the entire US economy.

Durbin and thirty-two other senators sponsored the DREAM Act. Though proposed several times beginning in 2001, this act has not yet passed Congress. The legislation would transform the lives of these young people, make better use of their energies and talents and, as studies show, provide a net benefit to all Americans.

President Obama created the Deferred Action for Childhood Arrivals (DACA) program in 2012 by executive order because opponents in Congress stalled the DREAM Act. This policy protects people brought into the United States as children, the same as the DREAM Act. DACA allows the DREAMers, people protected by this order, to remain in the US and apply for a work permit, a driver's license, and a Social Security number. It does not grant them any official status or a pathway to citizenship.

Immigration hard-liners oppose DACA and the DREAM Act. The Donald Trump administration worked to end DACA and stopped accepting applications. President Joe Biden reactivated the program on his first day in office on January 20, 2021. Biden plans to introduce sweeping immigration reform. His bill, the US Citizenship Act of 2021, provides an eight-year path to citizenship for almost all the estimated eleven million undocumented immigrants living in the US.

After basic training, the army sent Oscar to Afghanistan, where he served two tours of duty with distinction. He became a paratrooper and a ranger, attaining the rank of sergeant and leading his team in combat. Following his service, Oscar went to work for BNSF, a train company, as a supervisor in the machine shop. Oscar then moved into a position as a data analyst, working with statistical models, and he continues to pursue his goals.

"Ours is not necessarily an individual story, it's a team story," says Oscar. "Hardly ever does anyone do something entirely by themselves. There's always someone else." Then he adds, "Looking back, I can definitely see the influence that our teachers had on us."

Meanwhile, back at Carl Hayden Community High School, the 2004 robotics team created a legacy that the school and the community recognize. This team had become an inspiration to Carl Hayden students and the immigrant community. Under the coaching of Fredi and Allan, the robotics team continued to develop generations of engineering talent.

Ours is not necessarily an individual story, it's a team story. Hardly ever does anyone do something entirely by themselves. There's always someone else.

—*Oscar Vasquez*

The robotics team has sent more students to college with scholarship awards than all the athletics teams combined. Business leaders set up foundations to send students from Fredi and Allan's robotics teams to college. Says Peter Gaskins, one of these leaders, "Our country cannot afford to squander the talents of these kids. I'm just not willing to accept that this is the way it has to be."

CONCLUSION

Writing the stories of these nine teen inventors and engineers has been a daily inspiration to me. I want to be more like each of them. They are curious people who wonder why and then use their imaginations to wonder, "What if?"

> If there's a book that you want to read, but it hasn't been written yet, then you must write it.
>
> —*Toni Morrison*

Sometimes remarkable achievements seem mysterious and unattainable. These stories of invention are still remarkable and awesome, though less mysterious as you follow the process. Watching these developments unfold, you see that good and great ideas are absolutely attainable by mere mortals. Anyone can learn and use creative thinking to improve their skills.

Books in a school library opened a whole new world to William Kamkwamba. If these stories of young innovators and their inventions find their way into classrooms, homes, backpacks, and school libraries, will others find the confidence to try out a wacky idea? These teenagers experimented with untested ideas and ended up inventing ingenious devices. Along the way, and along with hard work, they learned a great deal and had fun. You could too. Maybe reading their stories will move you to begin a new chapter of your own story.

Global Challenges

The world needs people who are skilled in science and technology. In her final presentation for the Discovery Education 3M Young Scientist Challenge, Deepika Kurup referred to a set of fourteen grand global engineering challenges identified by the National Academy of Engineers (NAE). While we've made progress on many of these grand challenges, much more work remains.

These are the grand challenges:

- advance personalized learning
- make solar energy economical
- enhance virtual reality
- reverse-engineer the brain
- engineer better medicine
- advance health informatics
- restore and improve urban infrastructure
- secure cyberspace
- provide access to clean water
- provide energy from fusion
- prevent nuclear terrorism
- manage the nitrogen cycle
- develop carbon sequestration methods
- engineer the tools of scientific discovery

> You could change the world . . . just imagine what you could do.
>
> —*Jack Andraka*

Time magazine has just announced that it is convening a panel of innovative thinkers, which includes Gitanjali Rao, and a series of discussions focusing on the United Nations' Seventeen Sustainable Development Goals. Some of the UN goals not on the grand challenges list include these:

- ending poverty
- conquering AIDS, COVID-19, and other global health issues
- providing health, education, and safety for all children
- promoting human rights, independence, and democratic principles

- supporting global economic and social development
- resolving issues around population and migration
- securing world peace and supporting refugees

Over the next decade, *Time* intends to monitor and report on solving the challenges of a post-COVID-19 world. STEM skills will directly contribute to solving many of the challenges identified by the NAE and United Nations, such as clean water, affordable energy, solving climate change, and improved health. Other challenges, such as ending hunger and ensuring access to quality education for all, will require some combination of STEM skills, innovative ideas, leadership action, and collaborative redesign of our social and economic systems.

Deepika tells how she learned that solving global problems like these are not purely technical issues amenable to a quick fix. The people who solve global issues like these combine compassion, collaboration, and creative problem-solving methods, such as design thinking, along with STEM skills. William, when he worked for the design firm IDEO, used these design skills to work with communities in India to improve sanitation and in Kenya to help prevent violence against women. One of Jack Andraka's recent projects was to use cheap but effective biosensors he designed along with crowdsourcing data collection. These technologies track industrial pollution that contaminates drinking water in Tanzania back to the corporate culprits. Activists dip the sensors in a body of water, snap a picture with their cell phone, and upload it to a website to get a contamination reading. By tracking the flow of contaminants back to the source, these activists can find the polluters and hold them accountable.

Working on these global challenges is both exciting and meaningful. The world needs many more people working collectively on these challenges to bring about a better world.

As I write this, the SARS-CoV-2 coronavirus has ignited a global pandemic, killing over eight hundred thousand people in the US and five million globally. Anika Chebrolu, a fourteen-year-old scientist from Texas, recently won the 2020 Discovery Education 3M Young Scientist Challenge for her design of a potential antiviral drug that might work on COVID-19, as well as on the flu and other viral infections. Using computational methods, she found a compound that can bind to SARS-CoV-2, changing its shape and thus preventing the virus from infecting cells. While she stresses that her potential drug is still in the early stages of development, her approach is very promising. Anika was inspired to create this drug when she started looking at the problems in the world around her and wanted to use her knowledge of science to help solve them.

These lists of global challenges show we need everyone to solve these many daunting and complex problems. We cannot allow the talents of anyone to go to waste because we have stereotypes about who can be a scientist or an engineer. The stories in this book show that gender, age, sexual orientation, economic circumstance, or place of birth are not absolute barriers to discovery, invention, or creativity. If we as a society actively knock down these barriers, we will unleash torrents of ideas and energy. Everyone deserves the opportunity to develop and use their talents, and these collective talents are natural resources we cannot afford to waste.

Next Steps

We often assume creativity to be a gift that you either have or don't have. This is a myth. Work in neuroscience shows that creativity is both a habit and a skill. We can develop both habits and skills with practice—by just doing it.

If the idea of inventing devices to help people or designing better ways to do things excites you, then just go for it. Anyone can get started on an invention by reading and making.

Reading is a great way to get ideas and to learn how to do things. All of our teen innovators read and researched in books, in magazines, and on the web. Paraphrasing Jack, Google and Wikipedia are an inventor's two best friends. But don't spend all your time reading and learning. Start making things as soon as possible. Learn by doing.

Teens have tons of opportunities to get involved in STEM organizations and competitions. Many schools offer robotics teams as an extracurricular activity. Check out what your school has to offer and give it a try!

All of our teen inventors are makers. They all made things with their hands, learning as they went along. For an inventor, what you can build with your hands is much more important than what you can remember from a textbook for a multiple-choice test. Noted inventor Temple Grandin wrote a book for preteens on creativity and invention entitled *Calling All Minds*. She summarizes her entire message as "go make things," and the rest of her book shows how.

You can get ideas of what and how to make things at hands-on science museums, such as the Exploratorium in San Francisco and the Museum of Science in Boston. Go to a Maker Faire near you. Join a FIRST robotics team. Get an Arduino starter kit. Make some devices from the Exploratorium's Science Snacks website. Or just gather odds and ends from around your home—cardboard, tape, buttons, scraps of wood, old pieces of cloth, never-used items from your junk drawer—and begin creating a device you've imagined. These prototypes don't need to work. Leonardo da Vinci often began by sketching and then building nonworking prototypes of his fantastic ideas. Then he made them—at least the ones that were possible to make with the materials and tools of his time.

If you found these stories inspiring, create your own story—in STEM, the arts, social justice, or any other field that fires your imagination and drive.

Above all, have fun. That is the surest path to creative ideas.

GLOSSARY

aeration: purifying water by exposing it to the oxygen in air. Water treatment plants may spray the water into the air like a fountain or bubble air through the water. The oxygen in air helps remove dissolved gases, such as carbon dioxide, and oxidizes dissolved metals, such as iron and hydrogen sulfide, as well as volatile organic chemicals.

alternating current (AC): electrical current that flows first in one direction, then in the other, changing several times a minute. A power company can transmit alternating current for great distances along a power grid more easily than direct current (DC). Most of the electrical devices we plug into a wall socket use alternating current.

Alzheimer's disease: a disease that often results in the loss of memory, complex thinking abilities, and social skills. Advanced cases are serious enough to interfere with daily life. People who suffer from this disease may require special care. Scientists do not yet understand the causes. A number of factors contribute to this cognitive decline. These include changes in the brain as it ages as well as genetic, environmental, and lifestyle factors. Dementia is a more general term for the condition of people who suffer from the above symptoms. Not all dementia is Alzheimer's disease.

anthropology: the study of human cultures, evolution, and development

antibody: a protein created by the body to kill bacteria, viruses, and other microorganisms that can cause disease. Our immune system produces specific antibodies for each specific threat to our health.

appropriate technology: a movement in which designers create simple, useful devices that help people solve their daily problems by creatively using the tools and materials available in the community

Arduino: a company that produces kits and microcontroller boards for building digital devices that help students without a background in electronics or programming learn the basics of these skills. The term also refers to the computer chips, software, and kits the company makes.

bawo: a version of mancala, a strategy board game dating back to ancient Egypt. Two players take turns moving their markers around the board trying to capture the markers of their opponent.

biomarker: a substance in the body that may reveal a disease such as cancer

Bluetooth: technology that allows digital devices to communicate wirelessly over short distances using radio waves. For example, Bluetooth allows you to link your smartphone to wireless earbuds.

breadboard: a plastic board full of small holes that allows for easy insertion of electronic components and building of circuits

carbon nanotube: a molecule of carbon atoms that resemble a roll of chain-link fencing with six-sided links. Each nanotube is one hundred times stronger than steel and a better conductor of electricity than copper wire.

cassette player: a music-playing device about the size of a small radio commonly used before smartphones and MP3 players. These devices hold plastic cases, called cassettes, with magnetic tape wound around two spindles. The player moves the tape across magnetic surfaces that play the recorded music.

catalyst: any substance that speeds up a chemical reaction but is not part of the reaction. Catalysts remain unchanged by the reaction. Photocatalysts are catalysts that rely on light to speed up chemical reactions.

census: a count of the number of members of a population

Chichewa: one of the official languages of Malawi. It is part of the Bantu family of languages and is also called Chewa.

cholera: a deadly bacterial disease caused by eating or drinking contaminated food or water. People sick with cholera have severe vomiting and diarrhea. Cholera may be fatal if not treated right away. Clean, safe water supplies and proper community sanitation facilities stop the spread of cholera.

circuit board: a flat plastic sheet with circuit paths to connect the components of an electronic device. The "wires" are thin layers of copper foil attached to the board with heat and glue. Circuit boards are critical to electronic devices.

circuit breaker: a switch that automatically disconnects an electrical circuit when it overloads. It prevents fires by shutting off overheating circuits. Circuits overload when too much electrical current is flowing through them. Circuit breakers are like fuses, but breakers can be reset instead of replaced.

computational methods: a newer field of study that includes mathematical modeling, computer simulation, numerical analysis, and computer science theory

crochet: a craft similar to knitting used to make clothes or other items from yarn

Deferred Action for Childhood Arrivals (DACA): a policy that protects people who came to the United States illegally as children with their parents. DACA does not grant them any official status or a pathway to citizenship. It allows them to apply for a work permit, a driver's license, and a Social Security number. DACA covers about eight hundred thousand people, known colloquially as DREAMers.

deoxyribonucleic acid (DNA): a long, chainlike molecule that has a double-helix shape and contains the genetic code for building the entire organism. It is in every cell of every living organism.

deregulation: the removal of governmental control over the prices stores can charge for certain products

design thinking: a problem-solving approach in which designers work with communities to define issues and to brainstorm many potential solutions. The designers and the community try out the most promising ideas and revise them until they reach a good solution. Designing a new generation of computer chips is a technical problem. Designing a new system to improve service in a community clinic is a design-thinking problem.

Development, Relief, and Education for Alien Minors (DREAM) Act: a proposed policy that would provide a path to citizenship for young people who came to the United States illegally as children with their parents

direct current (DC): electric current that flows only in one direction, unlike alternating current (AC). Direct current from batteries supports portable devices and is not tied to the electric grid or wall sockets. Cars, flashlights, laptops, and headphones all use direct current.

dissipate: to reduce or dilute something

dynamo: a device that converts mechanical energy into electricity. For example, small dynamos attached to the back wheel of a bicycle can make the direct current to power a lamp.

dysentery: an infection of the intestines that causes severe diarrhea. Poor sanitation helps spread dysentery among people. Dysentery kills over half a million people each year, many of them babies.

electrical current: the flow of electrons through a material. These small, negatively charged particles move from one atom to another when pushed by a battery or a generator.

electrical resistance: a property of all materials that affects how easily electrons can flow through the material. The harder it is for a generator or a battery to push electrons through a material, the higher the electrical resistance of that material.

electromagnetic induction: the creation of electricity by moving a coil of wires through a magnetic field. If the wires are part of a closed circuit, electric current will flow through the circuit. The magnetic field produces the push to move the electrons. Electrical generators use this principle.

epidemiology: the study of the patterns and causes of disease and public health hazards. This research helps people understand risks to health and the precautions to take to prevent the spread of disease. Epidemiologists study events like the lead contamination in Flint, Michigan.

generator: a machine that produces electricity by rotating wires through a magnetic field to create a current in the wires

genetics: a branch of biology that studies DNA, genes, and the traits of an organism. Geneticists study how parents pass their traits to their offspring.

green card: also known as a permanent resident card, a document that grants someone permanent residency in the United States, allowing them to live and work here. A green card does not confer citizenship.

greenroom: a space where presenters or performers can relax before or after going onstage

homophobe: someone who has irrational fear, hatred, or mistrust of people who are gay, lesbian, or bisexual. Homophobia, a learned attitude, can include negative views and prejudiced behavior. Homophobes may bully, abuse, or violently injure people.

hovercraft: a vehicle that rides on a cushion of air produced by blowers and propelled by large fans. Hovercraft range from small tabletop models to the large craft used by the military to carry people and materials. Hovercraft can travel over land, water, mud, ice, swampland, or other surfaces.

immune system: the complex network of cells, tissues, and organs that work together to combat bacteria or viruses, defending the body against disease and infection

immunocomplex: a molecule formed when an antibody attaches onto an invading disease-causing microorganism. The antibody then calls on other parts of the immune system to eliminate this invading pathogen.

incubator: a small, low-temperature oven used in biology labs to grow cells under a steady temperature and constant conditions

index of refraction: the measure of how much a beam of light is bent when it passes from one medium to another, such as from air to water. For example, because of the index of refraction, a pencil partially below the surface in a glass of water looks bent from a side view.

induction current: an electrical current produced when a conductor, such as a wire, enters a moving magnetic field. Every wire with a flowing current has a magnetic field surrounding it.

infrastructure: networks of public projects such as roads, power plants, sewer lines, water treatment plants, and phone lines that help a community live better. Malawi and rural India are still building infrastructure, and the US is beginning to repair its aging infrastructure.

insulin: a chemical that regulates the amount of sugar in the blood. The pancreas produces this chemical to help the body use the energy from food. Pancreatic cancer interferes with this process. When a person's blood sugar level is too high, a doctor may diagnose them with diabetes and prescribe insulin injections.

iterative: using repetitive operations that build on one another

Junior Reserve Officer Training Corps (JROTC): a high school program sponsored by the US Department of Defense that encourages students to learn about the military and consider military service as a career

kerosene: an oil used for lighting when electric lights are not available. Kerosene lamps are smoky and can be dangerous in flammable houses.

kinetic energy: the energy of moving objects

kwacha: the currency of Malawi. As of November 2021, one US dollar is worth about 816 kwachas.

liquid crystal display (LCD): an electronic component using liquid crystals, a type of matter that flows like a liquid but has a structure like that of a crystal, to produce an image when lit from behind by a light source. These LCD displays are thinner and use less power than older light-emitting diode (LED) technology.

magnet school: a public school with special programs to attract students from across the district. These special programs may be in the performing arts, a specific career, or STEM fields.

maize: corn

mesothelin: a protein found on the surface of some cells, including cancer cells. Several types of cancer, including pancreatic tumors, cause the body to produce high levels of mesothelin.

metastasize: when cancer cells break away from the tumor and spread through the body. These cancer cells travel through the bloodstream or the lymph system and form new cancer tumors in other parts of the body. Once a cancer has metastasized, treating it is much more difficult.

microcontroller: a small electronic component that contains all the functions of a central processing unit of a computer. Often called a "computer on a chip," microcontrollers perform specific tasks. For example, a microcontroller in a smoke detector gets a signal from a smoke sensor and sends a signal that sounds an alarm. Many common electronic devices, from mobile phones to cars to microwaves to robots, use microcontrollers because they are so compact.

microelectronics: the use of tiny electronic components to create devices. Microcontrollers and liquid crystal displays (LCDs) are good examples of such components. Many modern devices, from computers to microwave ovens to satellites, are possible because of microcontrollers.

microorganism: a living thing too small for the unaided human eye to see. Most microorganisms, including many varieties of bacteria, viruses, and fungi, are harmless to humans, though some can make people sick.

mutate: to change, often referring to changes in the genetic material in a cell. Mutations can be harmful, helpful, or have no effect. The human ability to see a wide range of colors is a helpful mutation that most other mammals lack. The genetic disease cystic fibrosis results from a harmful mutation. Most mutations in the body have no effect.

nanoparticle: a piece of matter so small you need an electron microscope to see it. They range in size from about one billionth to one hundred billionths of a meter in diameter. Atoms and carbon nanotubes are examples of nanoparticles. Deepika used nanoparticles of titanium dioxide to purify water.

ohmmeter: a device that measures the force needed for electrical current to flow through a circuit

opioid: a powerful narcotic drug used to reduce a patient's pain. Oxycodone, hydrocodone, and codeine are some examples of opioids. Opioids are highly addictive. Opioid addiction has become a very serious problem. Overdoses of opioids can cause a person to pass out or stop breathing and die.

organic light-emitting diode (OLED): a small, flat display that lights up when electricity runs through special carbon material placed between thin film. This thin film is considered organic since it is a carbon compound. A diode is an electronic component that only allows current to flow in one direction. OLEDs are more efficient and much thinner than LCDs.

ozone: a colorless, unstable, toxic gas. It has a pungent odor and powerful oxidizing properties. Ozone forms when electrical discharges or ultraviolet light passes through oxygen. In our atmosphere, both solar radiation and lightning make ozone. A molecule of ozone has three atoms of oxygen (O_3). Most molecules of oxygen (O_2) have only two. Ozone high in the atmosphere protects Earth from excessive ultraviolet rays. Ozone at ground level contributes to smog and may aggravate asthma and other respiratory health issues.

pancreatic cancer: a disease caused by mutated cells that multiply rapidly and drain the body's resources that begins in the pancreas, an organ that produces enzymes to help digest food and hormones to manage blood sugar. Cancer cells may destroy normal cells and damage healthy tissue. Pancreatic cancer kills most of the people who have it.

parabolic reflector: a curved mirror that focuses light rays on a single point

pathogen: a bacterium, virus, or other microorganism that can cause disease

pervious: allowing water to flow through a material. Sand is very pervious, as water runs through it quickly.

polio: short for poliomyelitis, an infectious viral disease that attacks the central nervous system. Polio can cause temporary or permanent paralysis. The polio vaccine has nearly eliminated this disease in most parts of the world, according to the WHO.

polyvinyl chloride (PVC): a tough, flexible plastic with many uses, including plumbing

protein: a large, complex molecule that the body uses to build and repair tissues. Proteins also make the hormones, enzymes, and other chemicals the body needs. The body makes proteins from food.

prototype: an early test version of a device that allows the inventor to explore ideas, try out various functions, demonstrate the overall concept, and make improvements to the final device

pulse-width modulation (PWM): a type of digital signal to vary the average power delivered to a device. A small switch turns the current to the device on and off rapidly.

radiator: a device used to cool engines and prevent damage from overheating

radio transmitter: a device that turns voices or music into radio waves and then broadcasts these radio waves through the air. It can also convert radio waves back into voices or music.

reactive oxygen species (ROS): an unstable molecule containing oxygen that reacts easily with living cells. Natural metabolic processes in the body produce ROS. The immune system uses ROS to kill harmful microbes. Too many ROS molecules in a living cell may damage the DNA and other proteins and kill the cell. Exposure to toxins in the environment can produce harmful levels of ROS in humans.

remotely operated vehicle (ROV): a type of underwater robot connected to a tether or cable and operated by people on land or on a ship. ROVs are unoccupied and highly maneuverable. We use ROVs for underwater rescue, recovery, and exploration.

resistor: a small device used to slow or reduce the flow of current in a circuit by adding electrical resistance. William used a resistor to absorb some of the voltage his windmill produced so the circuit would not overheat.

rotor: the part of an electrical machine that rotates

sensor: a small electronic device that acts like our own senses to detect or measure something in the surrounding environment. There are sensors for motion, temperature, electrical resistance, acid levels, and many other uses.

signal processor: a chip that converts digitized data into a format used by its device. For example, the signal processor in a mobile phone compresses digital audio data.

silt: a material composed of very fine particles, smaller than sand and larger than clay. Silt particles are so small and light that they can stay suspended in water for a long time before settling at the bottom.

soldering iron: a tool used to melt solder, the soft metal "glue" that holds wires in place on a circuit board

sprocket: a toothed wheel that holds a bike chain in place as it spins. A bigger front sprocket is fixed to the pedals and crankshaft, and a small back sprocket is attached to the rear wheel. The sprockets are key parts of the drive mechanism of a bike.

subsistence farmer: a person who grows only enough food to get by

syringe: a small tube with a plunger that can take in or eject a thin stream of fluid

torque: a force that causes an object to twist or turn on an axis. For example, pushing a door open uses torque as the door rotates on its hinges.

tuition: money paid to attend school

tumor: a mass formed by a clump of rapidly growing cells. Some tumors are made of cancer cells, but not all. Doctors look for tumors during physical exams, but pancreatic cancer tumors are so deep in the body that doctors rarely find them in the early stages.

turbine: a large propeller that, when rotated by moving wind or water, drives a generator to produce electricity

typhoid: a bacterial infection of the body caused by eating contaminated food, drinking contaminated water, or having close contact with infected people. Sufferers may get headaches, stomach pains, nausea, or a high fever. Antibiotics can cure typhoid.

ultraviolet (UV): a form of electromagnetic radiation that falls between visible violet light and X-rays on the electromagnetic spectrum. Too much exposure to the high-energy UV rays from the sun may damage human tissues. A sunburn is an example of UV rays killing skin cells. These rays can also kill pathogens.

venture capitalist: a person who invests money in a new business to fund its growth in exchange for a share in the business's stock. Many new technologies and business ideas begin with venture funding. While many technology start-ups fail, successful companies, like Apple, pay very large returns.

volt: the unit of measurement of the electrical pressure in a circuit

voltage: the force, or "push," that causes electrical charges to move in a wire or other electrical conductor, thus causing an electrical current. We can compare voltage to water pressure in a plumbing system.

washer: a small disc of rubber or metal with a hole in the middle

SOURCE NOTES

13 Gitanjali Rao, personal interview with the author via web conferencing, April 16, 2020.

14 "A Promising Test for Pancreatic Cancer . . . from a Teenager: Jack Andraka," YouTube video, 10:49, posted by TED, July 11, 2013, https://www.youtube.com/watch?v=g-ycQufrgK4.

19 Jack Andraka with Matthew Lysiak, *Breakthrough: How One Teen Innovator Is Changing the World* (New York: HarperCollins, 2015), 60.

20 Andraka and Lysiak, 51.

20 Andraka and Lysiak, 41–42.

20 Andraka and Lysiak, 56.

21 Andraka and Lysiak, 76.

21 Andraka and Lysiak, 78.

21 Andraka and Lysiak, 78.

21–22 Andraka and Lysiak, 81.

22 Andraka and Lysiak, 87.

22 Andraka and Lysiak, 87.

22 Andraka and Lysiak, 88.

25 Andraka and Lysiak, 96.

25 Andraka and Lysiak, 97.

26 Andraka and Lysiak, 104.

26 Andraka and Lysiak, 106.

27 Andraka and Lysiak, 108.

27 Andraka and Lysiak, 108.

28 Andraka and Lysiak, 109.

29 Andraka and Lysiak, 110.

31 Andraka and Lysiak, 120.

32 Andraka and Lysiak, 121.

32 Andraka and Lysiak, 122.

32 Andraka and Lysiak, 125.

32 Andraka and Lysiak, 128.

33 "Thomas Edison," National Park Service, July 8, 2016, https://www.nps.gov/edis/learn/kidsyouth/timeline-of-edison-and-his-inventions.htm.

33 Andraka and Lysiak, *Breakthrough*, 132.

33 Andraka and Lysiak, 132.

33 Andraka and Lysiak, 132.

34 Andraka and Lysiak, 132.

34 Andraka and Lysiak, 136.

34 Andraka and Lysiak, 136.

34 Andraka and Lysiak, 137.

34 Andraka and Lysiak, 137.

35 Andraka and Lysiak, 140.

35 Andraka and Lysiak, 140.

37 Andraka and Lysiak, 149.

37 Andraka and Lysiak, 152.

37 Andraka and Lysiak, 154.

37 Andraka and Lysiak, 154.

37 Andraka and Lysiak, 154.

37–38 Andraka and Lysiak, 156.

38 Andraka and Lysiak, 172.

39 Nick Bryant, "The Prodigy Invention," *Sydney Morning Herald*, May 28, 2015, https://www.smh.com.au/lifestyle/the-prodigy-invention-20150527-ghax2d.html.

41 "2017 Final Presentations Gitanjali Rao," YouTube video, 10:41, posted by 3M Young Scientist Challenge, November 3, 2017, https://www.youtube.com/watch?v=5Cp8u6zszDQ.

41–42 Gitanjali Rao, personal interview with the author via web conferencing, November 3, 2019.

42 Rao.

42 Heather Mason, "America's Top Young Scientist Gitanjali Rao Talks about How Science Can Make a Difference," Amy Poehler's Smart Girls, October 30, 2017, https://amysmartgirls.com/americas-top-young-scientist-gitanjali-rao-talks-about-how-science-can-make-a-difference-f337e51a9504/.

43 Sara Ganim, "5,300 U.S. Water Systems Are in Violation of Lead Rules," CNN, June 29, 2016, https://www.cnn.com/2016/06/28/us/epa-lead-in-u-s-water-systems/index.html.

44 Rao, personal interview, April 16, 2020.

45 Rao.

46 Jacopo Prisco, "Gitanjali Rao Wants to Make Water Safer with Lead Detection System," CNN, February 18, 2018, https://www.cnn.com/2017/11/28/health/gitanjali-rao-young-scientist-winner/index.html.

46 Rao, personal interview, November 3, 2019.

46 Rao, personal interview, April 16, 2020.

46 Rao.

46 Rao.

47 Rao.

47 Rao.

47 Katie Kindelan, "11-Year-Old Girl Inspired by Flint Water Crisis Creates Cheap Kit to Test Lead," ABC News, October 18, 2017, https://abcnews.go.com/Lifestyle/11-year-girl-inspired-flint-water-crisis-creates/story?id=50559884.

47 Rao, personal interview, November 3, 2019.

47–48 Rao, personal interview, April 16, 2020.

48 Rao.

49 Kelly Hall, "11-Year-Old Scientist Is Developing a Solution to Help Solve the Water Crisis in Flint, Michigan," 3M, accessed July 9, 2020, https://www.3m.com/3M/en_US/particles/all-articles/article-detail/~/clean-water-lead-detection-young-scientist-challenge/?storyid=e8ea94e9-95b7-448a-bd74-79fbbb0a5960/.

49 Rose Lichter-Marck, "Finding Solutions to Real Problems: An Interview with Gitanjali Rao," Rookie, January 11, 2018, https://www.rookiemag.com/2018/01/finding-solutions-real-problems-interview-gitanjali-rao/.

49 Rao, personal interview, April 16, 2020.

49 Rao, personal interview, November 3, 2019.

49 Rao, personal interview, April 16, 2020.

50 Lichter-Marck, "Finding Solutions."

50 Hall, "11-Year-Old Scientist."

50 Rao, personal interview, April 16, 2020.

50 Hall, "11-Year-Old Scientist."

50 "2017 National Finalist: Gitanjali Rao," YouTube video, 1:59, posted by 3M Young Scientist Challenge, July 18, 2017, https://www.youtube.com/watch?v=m4WM3arrBgo/.

50–51 Rao, personal interview, April 16, 2020.

51 Rao.

51 Katie Harveston, "5 Amazing Inventions by Young People," Techmalak, January 19, 2018, https://techmalak.com/5-amazing-tech-inventions-by-young-people/#.WmX9QJM-fR0.

51 Hall, "11-Year-Old Scientist."

51–52 Harry Smith, "Meet the 13-Year-Old Genius on a Limitless Quest for Knowledge," *Today*, November 4, 2018, https://www.today.com/video/meet-the-13-year-old-genius-on-a-limitless-quest-for-knowledge-1361509443705.

52 Hall, "11-Year-Old Scientist."

52 Adriana Diaz, "12-Year-Old Colorado Girl, Troubled by Flint Water Crisis More Than a 1,000 Miles Away, Invents Lead Detector," CBS News, December 25, 2017, https://www.cbsnews.com/news/gitanjali-rao-12-year-old-girl-troubled-by-flint-water-crisis-invents-lead-detector/.

52 Hall, "11-Year-Old Scientist."

52 Rao, personal interview, April 16, 2020.

52 Hall, "11-Year-Old Scientist."

52 Rao, personal interview, April 16, 2020.

54 Rao.

54 Hall, "11-Year-Old Scientist."

54 Hall.

54 Foluké Tuakli, "How Flint's Fight for Clean Water Inspired 'America's Top Young Scientist,'" NBC News, October 31, 2017, https://www.nbcnews.com/news/asian-america/how-flint-s-fight-clean-water-inspired-america-s-top-n813966.

55 Kindelan, "11-Year-Old Girl Inspired by Flint Water Crisis."

55 Diaz, "12-Year-Old Colorado Girl."

55 Rao, personal interview, April 16, 2020.

55 Hall, "11-Year-Old Scientist."

56 Michael Elizabeth Sakas, "13-Year-Old Gitanjali Rao Lead Detecting Invention Lands Her on Forbes' '30 under 30,'" CPR News, January 28, 2019, https://www.cpr.org/2019/01/28/13-year-old-gitanjali-raos-lead-detecting-invention-lands-her-on-forbes-30-under-30/.

56 Rao, personal interview, April 16, 2020.

57 Nick Puckett, "Lone Tree Teen Takes on Opioid Epidemic," Lone Tree (CO) Voice, July 30, 2019, https://lonetreevoice.net/stories/lone-tree-teen-takes-on-opioid-epidemic,284178.

57 Rao, personal interview, April 16, 2020.

57 Rao.

58 Rao, personal interview, November 3, 2019.

58 Smith, "Meet the 13-Year-Old Genius."

58 Rao, personal interview, November 3, 2019.

59 Rao, personal interview, April 16, 2020.

59 Abigail Hess, "This 12-Year-Old Won $25,000 for an Invention That Helps Detect Lead in Water," CNBC, January 10, 2018, https://www.cnbc.com/2018/01/10/this-12-year-old-won-25000-for-a-gadget-that-helps-detect-lead.html.

59 Time Staff, "Meet TIME's First-Ever Kid-of-the-Year," Time, December 3, 2020, https://time.com/5916772/kid-of-the-year-2020/.

59 Hess, "This 12-Year-Old Won."

61 William Kamkwamba, "How I Harnessed the Wind," TED, July 2009, https://www.ted.com/talks/william_kamkwamba_how_i_harnessed_the_wind?language=en/.

61 William Kamkwamba and Brian Mealer, *The Boy Who Harnessed the Wind: A True Story of Survival against the Odds* (New York: Penguin, 2015), 6.

61–62 Kamkwamba and Mealer, 6.

63 Kamkwamba and Mealer, 67–69.

63 Kamkwamba and Mealer, 67–69.

64 Kamkwamba and Mealer, 67–69.

64 Kamkwamba and Mealer, 79.

65 Kamkwamba and Mealer, 80.

65 Kamkwamba and Mealer, 81.

65 Kamkwamba and Mealer, 76.

66 Kamkwamba and Mealer, 121.

67 Kamkwamba and Mealer, 160.

68 Kamkwamba and Mealer, 165–166.

69 Kamkwamba and Mealer, 165–166.

69 Kamkwamba and Mealer, 165–166.

69–70 Kamkwamba and Mealer, 167–168.

70 Mary Atwater, *Using Energy* (New York: MacMillan McGraw Hill, 1995), 61.

70 Kamkwamba and Mealer, Boy, 167–168.

70 Kamkwamba and Mealer, 169.

70 Kamkwamba and Mealer, 169.

73 Kamkwamba and Mealer, 159.

73 Kamkwamba and Mealer, 175.

73 Kamkwamba and Mealer, 177.

74 Kamkwamba and Mealer, 183.

75 Kamkwamba and Mealer, 189.

75 Kamkwamba and Mealer, 198.

77–78 Kamkwamba and Mealer, 204.

78 Kamkwamba and Mealer, 204.

78 Kamkwamba and Mealer, 204.

79 Kamkwamba and Mealer, 204.

80 Kamkwamba and Mealer, 249.

80 Kamkwamba and Mealer, 249.

80 Kamkwamba and Mealer, 268.

81 Kamkwamba and Mealer, appendix, 4.

83 "Spring Meetings 2019 Global Voices: Interview with William Kamkwamba," Facebook video, 11:22, posted by World Bank, April 11, 2019, https://www.facebook.com/worldbank/videos/1046368292216402/.

83 Kamkwamba and Mealer, Boy, 285.

83 Kamkwamba, "How I Harnessed the Wind."

86 Austin Veseliza, personal interview with the author, September 8, 2016.

86–87 Veseliza, personal interview with the author via web video communication, July 2, 2020.

87 Veseliza.

87 Veseliza.

87–88 Veseliza, personal interview, September 8, 2016.

88 Veseliza, personal interview, July 2, 2020.

88 Veseliza.

88 Veseliza, personal interview, September 8, 2016.

91 Veseliza, personal interview, July 2, 2020.

93 Veseliza, personal interview, September 8, 2016.

93 Veseliza, personal interview, July 2, 2020.

95 Veseliza.

95 Veseliza.

96 Veseliza.

96 Veseliza.

99 Veseliza.

101 Veseliza, personal interview, September 8, 2016.

102 Veseliza, personal interview, July 2, 2020.

102 Veseliza.

102 Veseliza.

103 Veseliza.

104 Veseliza.

104 Veseliza.

104 Veseliza.

105 Veseliza.

105 Veseliza.

105 Veseliza.

105 Veseliza.

106 Veseliza.

106 Veseliza.

106 Veseliza.

109 Meghan Modafferi, "Deepika Kurup: Student Scientist," National Geographic, March 29, 2016, https://blog.education.nationalgeographic.org/2016/03/29/deepika-kurup-student-scientist/.

109 Deepika Kurup, "A Young Scientist's Quest for Clean Water," TED, October 2016, https://www.ted.com/talks/deepika_kurup_a_young_scientist_s_quest_for_clean_water.

109–110 Modafferi, "Deepika Kurup."

110 "Is Deepika Kurup the Most Talented Sophomore at Harvard?," Tab, accessed October 14, 2020, https://thetab.com/us/harvard/2015/12/24/is-deepika-kurup-the-most-talented-freshman-at-harvard-2016.

110 Modafferi, "Deepika Kurup."

110 Kurup, "A Young Scientist's Quest."

111 Kurup.

111 "Talented Sophomore," Tab.

111 Kurup, "A Young Scientist's Quest."

111 Modafferi, "Deepika Kurup."

111 Deepika Kurup, "A STEM Girl's Opportunities: From Science to the White House," Huffington Post, updated December 6, 2017, https://www.huffpost.com/entry/girls-in-stem_b_3275468.

112 Kurup.

112 Kurup.

112 Modafferi, "Deepika Kurup."

112 Deepika Kurup, "Photocatalytic Composition for Water Purification," Google Patents, accessed October 14, 2020, https://patents.google.com/patent/US20140183141A1/en.

114 Amy Crawford, "Interview: Amy Smith, Inventor," Smithsonian, September 2006, https://www.smithsonianmag.com/arts-culture/interview-amy-smith-inventor-130472633.

117 Kurup, "A Young Scientist's Quest."

117 Kurup.

117 Kurup.

118 "A Teen Is Taking on the Global Water Crisis," A Plus, August 20, 2018, https://aplus.com/v/85132/a-teen-is-taking-on-the-global-water-crisis/.

118 Deepika Kurup, "Photocatalytic Rod for Green and Sustainable Water Purification," youngscientistlab.com, accessed https://www.youngscientistlab.com/entry/352.

118 "A Teen Is Taking on the Global Water Crisis," A Plus.

118 Kurup, "Photocatalytic Rod."

118 David Abel, "Pollution Plagues the Mighty Merrimack When Rain Is Heavy," Merrimack River Watershed Council, April 21, 2020, https://merrimack.org/2020/04/21/pollution-plagues-the-mighty-merrimack/.

119 Kurup, "Photocatalytic Rod."

119 Kurup.

120 Kurup.

121 Jothi Ramaswamy, "Deepika Kurup Interview," thinksteam4girls.org, October 28, 2015, https://www.thinksteam4girls.org/deepika-kurup-interview/.

121 Ramaswamy.

123 Deepika Kurup, "Deepika FOX News Interview," YouTube video, 3:52, posted by Deepika Kurup, October 12, 2014, https://www.youtube.com/watch?v=LWvaKK_DL6A.

123 Kurup.
124 Kurup, "A STEM Girl's Opportunities."
124 Kurup.
124 Kurup.
126 Deepika Kurup, "A Novel Photocatalytic Pervious Composite for Degrading Organics and Inactivating Bacteria in Wastewater," Entry to the Stockholm Junior Water Prize 2014, 4.
126 "A Teen Is Taking on the Global Water Crisis," A Plus.
126 "Talented Sophomore," Tab.
126 "Fellows: Scholarship Categories," Davidson Institute, accessed September 13, 2021, https://www.davidsongifted.org/gifted-programs/fellows-scholarship/scholarship-categories/.
127 "Talented Sophomore," Tab.
127 "Talented Sophomore."
127 "Talented Sophomore."
127 "Talented Sophomore."
128 "A Teen Is Taking on the Global Water Crisis," A Plus.
128 "A Teen Is Taking on the Global Water Crisis."
128 "A Teen Is Taking on the Global Water Crisis."
128 Modafferi, "Deepika Kurup."
128 Deepika Kurup, "Last week was my white coat ceremony @StanfordMed!," Twitter post, August 31, 2019, https://twitter.com/TheDeepikaKurup/status/1167907842289876994/photo/1.
129 Kurup.
129 Pamme Boutselis, "Meet Deepika Kurup, 2014 TedxAmoskeagMillyard Speaker," TedxAmoskeagMillyard (blog), September 16, 2014, https://tedxamoskeagmillyard.wordpress.com/2014/09/16/meet-deepika-kurup-2014-tedxamoskeagmillyard-speaker/.
129 "Talented Sophomore," Tab.
129 "A Teen Is Taking on the Global Water Crisis," A Plus.
133 Joshua Davis, *Spare Parts: Four Undocumented Teenagers, One Ugly Robot, and the Battle for the American Dream* (New York: Farrar, Straus and Giroux, 2014), 96.
133 Davis, 20.
134 Davis, 96.
134 Davis, 96.
134 Davis, 96.
134 Davis, 96.
138 Davis, 84.
139 Davis, 86.
139 Davis, 93.
139 Davis, 97.
141 Davis, 75.
142 Davis, 98.
142 Davis, 98.
142 Davis, 100.
142 Davis, 100.
143 Davis, 100.
143 Davis, 100.
143 Davis, 100.
144 Davis, 104.

146 Davis, 112.

146 Davis, 112.

148 Davis, 124.

150 Davis, 138.

150 Davis, 138.

151 Davis, 138.

153 Davis, 157.

154 Davis, 5.

154 Davis, 5.

154 Davis, 5.

154 Davis, 5.

154 Davis, 5.

155 Davis, 161.

155 Davis, 162.

156 Davis, 165.

158 Davis, 171.

159 Davis, 172.

160 Davis, 187.

160 Davis, 187.

160 Davis, 189.

160 Davis, 189.

160 Davis, 189.

161 Davis, 187.

167 Davis, 189.

167 Davis, 189.

167 Davis, 189.

168 Christine Hult, *The Handy English Grammar Answer Book* (Canton, MI: Visible Ink, 2015), 327.

169 "A Promising Test for Pancreatic Cancer," YouTube video.

202 "Maker Faire: A Bit of History," Maker Faire, accessed September 15, 2021, https://makerfaire.com/makerfairehistory/.

SELECTED BIBLIOGRAPHY

Abel, David. "Pollution Plagues the Mighty Merrimack River When Rain Is Heavy." Merrimack River Watershed Council, April 21, 2020. http://merrimack.org/2020/04/21/pollution-plagues-the-mighty-merrimack/.

"AmeriCorps Group Comes from Ft. Stanton to Lend a Helping Hand to Historical Society." *Roswell (NM) Daily Record*, May 26, 2018. https://www.rdrnews.com/2018/05/26/americorps-group-comes-from-ft-stanton-to-lend-helping-hand-to-historical-society/.

"Ancient Animals." Leatherback Trust. Accessed July 13, 2020. https://www.leatherback.org/why-leatherbacks/ancient-animals/.

Anderson, Kent. "The Jack Andraka Story—Uncovering the Hidden Contradictions behind a Science Folk Hero." Scholarly Kitchen, January 3, 2014. https://scholarlykitchen.sspnet.org/2014/01/03/the-jack-andraka-story-uncovering-the-hidden-contradictions-of-an-oa-paragon/.

Andraka, Jack, with Matthew Lysiak. *Breakthrough: How One Teen Innovator Is Changing the World*. New York: HarperCollins, 2015.

"Boil Water Response—Information for the Public Health Professional." New York State Department of Health, November 2018. https://www.health.ny.gov/environmental/water/drinking/boilwater/response_information_public_health_professional.htm.

Brown, Tim. *Change by Design: How Design Thinking Transforms Organizations and Inspire Innovations*. New York: Harper Business, 2009.

Bryant, Nick. "The Prodigy Invention." *Sydney Morning Herald*, May 28, 2015. https://www.smh.com.au/lifestyle/the-prodigy-invention-20150527-ghax2d.html.

"Catalyst." *Encyclopedia Britannica*, August 15, 2019. https://www.britannica.com/science/catalyst/.

Chen, Guangzhe. "The Numbers Are In: Water Is Key to Poverty Reduction and Health." World Bank, March 27, 2017. https://blogs.worldbank.org/water/water-key-poverty-reduction-and-health/.

CNN Editorial Research. "Flint Water Crisis Fast Facts." CNN, December 13, 2019. https://www.cnn.com/2016/03/04/us/flint-water-crisis-fast-facts/index.html.

Davis, Joshua. "A Cruel Waste of America's Tech Talent." *New York Times*, January 15, 2015. https://www.nytimes.com/2015/01/16/opinion/the-cruel-waste-of-americas-tech-talent.html.

———. "La Vida Robot: How Four Underdogs from the Mean Streets of Phoenix Took on the Best from M.I.T. in the National Underwater Bot Championship." *Wired*, April 1, 2005. https://www.wired.com/2005/04/la-vida-robot/.

———. *Spare Parts: Four Undocumented Teenagers, One Ugly Robot, and the Battle for the American Dream*. New York: Farrar, Straus and Giroux, 2014.

DeRuy, Emily, and Geneva Sands. "The True Story of How 4 Undocumented Teens from Phoenix Beat MIT in a Robotics Competition." Splinter, January 16, 2015. https://splinternews.com/the-true-story-of-how-4-undocumented-teens-from-phoenix-1793844776/.

Diaz, Adriana. "12-year-old Colorado Girl, Troubled by Flint Water Crisis More Than 1,000 Miles Away, Invents Lead Detector." CBS News, December 25, 2017. https://www.cbsnews.com/news/gitanjali-rao-12-year-old-girl-troubled-by-flint-water-crisis-invents-lead-detector/.

"DNA Structure and Function." Khan Academy. Accessed October 29, 2020. https://www.khanacademy.org/test-prep/mcat/biomolecules/dna/a/dna-structure-and-function.

"The Economic Impact of S. 744, the Border Security, Economic Opportunity, and Immigration Modernization Act." Congressional Budget Office, June 18, 2013. http://www.cbo.gov/publication/44346.

"The Effects of Immigration on the U.S. Economy." Penn Wharton, University of Pennsylvania, July 27, 2016. https://budgetmodel.wharton.upenn.edu/issues/2016/1/27/the-effects-of-immigration-on-the-united-states-economy/.

Elassar, Alaa. "This 14-Year-Old Girl Won a $25K Prize for a Discovery That Could Lead to a Cure for COVID-19." CNN, October 18, 2020. https://www.cnn.com/2020/10/18/us/anika-chebrolu-covid-treatment-award-scn-trnd/index.html.

Ganim, Sara. "5,300 U.S. Water Systems Are in Violation of Lead Rules." CNN, June 29, 2016. https://www.cnn.com/2016/06/28/us/epa-lead-in-u-s-water-systems/index.html.

"Gitanjali Rao, Karlie Kloss, & Megan Smith: MAKERS Conference 2018." Yahoo News video, 9:05, February 6, 2018. https://www.yahoo.com/lifestyle/gitanjali-rao-karlie-kloss-megan-212613863.html.

Hall, Kelly. "11-year-old Scientist Is Developing a Solution to Help Solve the Water Crisis in Flint, Michigan." 3M. Accessed July 9, 2020. https://www.3m.com/3M/en_US/particles/all-articles/article-detail/~/clean-water-lead-detection-young-scientist-challenge/?storyid=e8ea94e9-95b7-448a-bd74-79fbbb0a5960/.

Harveston, Katie. "5 Amazing Inventions by Young People." Techmalak, January 19, 2018. https://techmalak.com/5-amazing-tech-inventions-by-young-people/#.WmX9QJM-fR0/.

Hess, Abigail. "This 12-Year-Old Won $25,000 for an Invention That Helps Detect Lead in Water." CNBC, January 10, 2018. https://www.cnbc.com/2018/01/10/this-12-year-old-won-25000-for-a-gadget-that-helps-detect-lead.html.

Kamkwamba, William. "Moving Windmills: The William Kamkwamba Story." YouTube video, 6:07. Posted by William Kamkwamba, February 14, 2008. https://www.youtube.com/watch?v=arD374MFk4w/.

———. "When It Comes to Climate Change, We Should Start Small, Fail Fast, and Dream Big." Huffington Post, December 6, 2017. https://www.huffpost.com/entry/when-it-comes-to-climate_b_9994172/.

Kamkwamba, William, and Brian Mealer. *The Boy Who Harnessed the Wind: A True Story of Survival against the Odds*. New York: Penguin, 2015.

Kindelan, Katie. "11-year-old Girl Inspired by Flint Water Crisis Creates Cheap Kit to Test Lead." ABC News, October 18, 2017. https://abcnews.go.com/Lifestyle/11-year-girl-inspired-flint-water-crisis-creates/story?id=50559884/.

Kurup, Deepika. "Deepika Kurup: Student, Scientist, Advocate." deepikakurup.com. Accessed October 20, 2020. http://deepikakurup.com/.

———. "Photocatalytic Composition for Water Purification." US Patent 2014/0183141 A1, filed December 30, 2013, and issued July 3, 2014. Accessed October 14, 2020. https://patents.google.com/patent/US20140183141A1/en/.

Lichter-Marck, Rose. "Finding Solutions to Real Problems: An Interview with Gitanjali Rao." Rookie, January 11, 2018. https://www.rookiemag.com/2018/01/finding-solutions-real-problems-interview-gitanjali-rao/.

Mason, Heather. "America's Top Young Scientist Gitanjali Rao Talks about How Science Can Make a Difference." Amy Poehler's Smart Girls, October 30, 2017. https://amysmartgirls.com/americas-top-young-scientist-gitanjali-rao-talks-about-how-science-can-make-a-difference-f337e51a9504/.

Newnham, Danielle. "The Story of Shazam: The Start Up Days." Medium.com, November 11, 2015. https://medium.com/swlh/the-story-of-shazam-the-startup-days-6bccebd17d84.

"1 in 3 People Globally Do Not Have Access to Safe Drinking Water." World Health Organization, June 18, 2019. https://www.who.int/news/item/18-06-2019-1-in-3-people-globally-do-not-have-access-to-safe-drinking-water-unicef-who/.

Prisco, Jacopo. "Gitanjali Rao Wants to Make Water Safer with Lead Detection System." CNN, February 18, 2018. https://www.cnn.com/2017/11/28/health/gitanjali-rao-young-scientist-winner/index.html.

"A Promising Test for Pancreatic Cancer . . . from a Teenager: Jack Andraka." YouTube video, 10:49. Posted by TED, July 11, 2013. https://www.youtube.com/watch?v=g-ycQufrgK4.

Puckett, Nick. "Lone Tree Teen Take on Opioid Epidemic." *Lone Tree (CO) Voice*, July 30, 2019. https://lonetreevoice.net/stories/lone-tree-teen-takes-on-opioid-epidemic,284178.

Ruelas, Richard. "10 Years Ago They Beat MIT. Today It's Complicated." AZ Central, July 17, 2014. https://www.azcentral.com/story/life/az-narratives/2014/07/17/phoenix-high-school-win-mit-resonates-decade-later/12777467/.

Sakas, Michael Elizabeth. "13-Year-Old Gitanjali Rao Lead Detecting Invention Lands Her on *Forbes*' '30 under 30.'" CPR News, January 28, 2019. https://www.cpr.org/2019/01/28/13-year-old-gitanjali-raos-lead-detecting-invention-lands-her-on-forbes-30-under-30/.

"Sea Turtle Evolution." Seaturtle World, January 4, 2014. https://www.seaturtle-world.com/sea-turtle-evolution/.

Smith, Harry. "Meet the 13-Year-Old Genius on a Limitless Quest for Knowledge." *Today*, November 4, 2018. https://www.today.com/video/meet-the-13-year-old-genius-on-a-limitless-quest-for-knowledge-1361509443705/.

Smith, Roff. "Here's What Happened the Day the Dinosaurs Died." *National Geographic*, June 11, 2016. https://www.nationalgeographic.com/news/2016/06/what-happened-day-dinosaurs-died-chicxulub-drilling-asteroid-science/.

"Solar Disinfection." Centers for Disease Control and Prevention, October 10, 2012. https://www.cdc.gov/safewater/solardisinfection.html.

"Solar Water Disinfection." Wikipedia. Accessed October 15, 2020. https://en.wikipedia.org/wiki/Solar_water_disinfection.

"Statistics on Voice, Language, and Speech." National Institutes of Health, National Institute on Deafness and Other Communication Disorders, July 11, 2016. https://www.nidcd.nih.gov/health/statistics/statistics-voice-speech-and-language/.

Stout, Robert Joe. *Why Immigrants Come to America: Braceros, Indocumentados, and the Migra*. Westport, CT: Greenwood, 2007.

Sullivan, Kathleen J. "Stanford Junior Wins 2018 Truman Scholarship for Graduate Studies." Stanford University, April 13, 2018. https://news.stanford.edu/2018/04/13/truman-scholar/.

Teel, Roger. "Army Awards Scholarships to eCYBERMISSION National Winners." US Army, June 28, 2010. https://www.army.mil/article/41558/army_awards_scholarships_to_ecybermission_national_winners/.

"TIME Announces Decade-Long 'TIME 2030' Project to Explore Solutions to the Challenges of a Post-COVID World." *Time*, January 25, 2021. https://time.com/5932958/time-2030-project-post-covid-world/.

Time Staff. "Meet *TIME*'s First-Ever Kid of the Year." *Time*, December 3, 2020. https://time.com/5916772/kid-of-the-year-2020/.

"The True Story of the Kids Who Beat MIT's Best Robots, Coming Soon to Theaters." *Wired*, December 2, 2014. https://www.wired.com/2014/12/spare-parts/.

Tuakli, Foluké. “How Flint’s Fight for Clean Water Inspired ‘America’s Top Young Scientist.’” NBC News, October 31, 2017. https://www.nbcnews.com/news/asian-america/how-flint-s-fight-clean-water-inspired-america-s-top-n813966/.

Tucker, Abigail. “Jack Andraka, the Teen Prodigy of Pancreatic Cancer.” Smithsonian, December 2012. https://www.smithsonianmag.com/science-nature/jack-andraka-the-teen-prodigy-of-pancreatic-cancer-135925809/.

“2017 Final Presentations—Gitanjali Rao.” YouTube video, 10:41. Posted by 3M Young Scientist Challenge, November 3, 2017. https://www.youtube.com/watch?v=5Cp8u6zszDQ/.

“2017 National Finalist: Gitanjali Rao.” YouTube video, 1:59. Posted by 3M Young Scientist Challenge, July 18, 2017. https://www.youtube.com/watch?v=m4WM3arrBgo/.

Upbin, Bruce. “Wait, Did This 15-Year-Old Kid from Maryland Just Change Cancer Treatment?” *Forbes*, June 18, 2012. https://www.forbes.com/sites/bruceupbin/2012/06/18/wait-did-this-15-year-old-from-maryland-just-change-cancer-treatment/.

Vargas, Jose Antonio. *Dear America: Notes of an Undocumented Citizen.* New York: HarperCollins, 2018.

Wagner, Kurt. “Silicon Valley Teens on the Future of Technology.” Vox, October 5, 2014. https://www.vox.com/2014/10/5/11631590/silicon-valley-teens-on-the-future-of-technology/.

“Water for Health—Taking Charge.” World Health Organization. Accessed October 14, 2020. https://www.who.int/water_sanitation_health/takingcharge.html.

West, Darrell. *Brain Gain: Re-Thinking U.S. Immigration Policy.* Washington, DC: Brookings Institution, 2010.

“What Causes a Sea Turtle to Be Born Male or Female?” National Ocean Service, April 9, 2020. https://oceanservice.noaa.gov/facts/temperature-dependent.html.

“Young Scientist Anika: Reimagining a Healthier World.” Facebook video, 8:12. Posted by UNICEF, November 19, 2020. https://www.facebook.com/unicef/videos/877821652986699/.

FURTHER INFORMATION

INTRODUCTION

Earthwatch
https://earthwatch.org
Earthwatch sends volunteers to aid with sea turtle research each year.

The Goldring-Gund Marine Biology Station
https://www.leatherback.org/conservation/research/goldring-gund-marine-biology-station/
This research arm of the Leatherback Trust hosts middle school and high school groups, like Martin and Daniel's, for field trips.

Instructables
https://www.instructables.com/
This how-to site has many projects you can do at home. They range from projects for beginners to advanced. For example, here's an advanced plan created by someone who wanted a supply of green energy for his remote cabin: https://www.instructables.com/How-I-built-an-electricity-producing-wind-turbine/. This is not a beginner's project. As William did, start small and simple before going big and complex. If you go to the menu along the top of the home page and click on "Teacher," you can select appropriate projects by grade and subject.

Spotila, James R. *Sea Turtles: A Complete Guide to Their Biology, Behavior, and Conservation.* Baltimore: Johns Hopkins University Press, 2004.

Spotila, James R., and Pilar Santidrian Tomillo, eds. *The Leatherback Turtle: Biology and Conservation.* Baltimore: Johns Hopkins University Press, 2015.
These two books by James Spotila, one of the founders of the Leatherback Trust, are a great place to learn so much more about these ancient and important animals.

CHAPTER 1

American Cancer Society
https://www.cancer.org/cancer/pancreatic-cancer/
Read about pancreatic cancer at this American Cancer Society site. The society sponsors cancer research, provides information on cancer, advocates for people with cancer, and raises funds.

National Suicide Prevention Lifeline
https://suicidepreventionlifeline.org/
This organization has crisis centers all over the United States and offers resources for learning more on preventing suicide. The phone lifeline, 800-273-8255, takes calls twenty-four hours a day, every day.

"A Promising Test for Pancreatic Cancer from a Teenager"
https://www.ted.com/talks/jack_andraka_a_promising_test_for_pancreatic_cancer_from_a_teenager?language=en/
Jack's compelling and inspiring TED Talk summarizes his quest to find a better way to detect pancreatic cancer earlier and save more lives, like that of Uncle Ted.

Stop Bullying
https://www.stopbullying.gov
This US Department of Health and Human Services website lists many resources for teachers and schools, parents and the public, and for young people. A section on laws and government policies relates to bullying as well as training on how to address bullying. This is a good source if you or someone you know is being bullied.

CHAPTER 2

Clark, Anna. *The Poisoned City: Flint's Water and the American Urban Tragedy*. New York: Macmillan, 2019.
Journalist Anna Clark's prize-winning book delves into the origins of the crisis and the subsequent mismanagement, poor decisions, deception, and cover-ups by officials and politicians. Clark relates the long history of the municipal government neglecting the citizens of Flint, 54 percent of whom are Black and 38 percent of whom fall under the US government poverty line.

Field, Simon Quellen. *Why There's Antifreeze in Your Toothpaste: The Chemistry of Household Ingredients*. Chicago: Chicago Review Press, 2007.
This absorbing book is by a science writer trained in biochemistry. Did you ever want to know what the strange-sounding ingredients with impossibly long chemical names were doing in your food or household products? Field explains what each chemical is, what its formula is, and why it's used. He covers preservatives, flavorings, sweeteners, food coloring, hair conditioner, and toothpaste, along with eighteen other types of chemicals in our daily lives.

"Flint Water Crisis: Everything You Need to Know"
https://www.nrdc.org/stories/flint-water-crisis-everything-you-need-know/
As the title says, this site has a concise history of the crisis and the key facts about the water crisis the people in Flint, Michigan, are suffering from.

Gitanjali Rao's Website
https://gitanjalirao.net/
Gitanjali is an influencer and a connector, and her sites list some of the partnerships and collaborations she has formed. You can sign up for one of Gitanjali's innovation workshops on this site and learn more about her current projects.

Hanna-Attisha, Mona. *What the Eyes Don't See: A Story of Crisis, Resistance, and Hope in an American City*. London: One World, 2019.
Dr. Hanna-Attisha is a pediatrician who helped discover and expose the horror of the Flint children and families exposed to toxic levels of lead in their drinking water. This is her story of her struggles to get officials and politicians to act to stop the harm she saw Flint's contaminated water doing to her patients.

Keen, Sam. *The Disappearing Spoon: And Other True Tales of Madness, Love, and the History of the World from the Periodic Table of the Elements*. New York: Little, Brown, 2010.
Keen's fascinating book is a great way to learn chemistry through the history of science rather than by memorizing the familiar wall chart. Keen makes the

periodic table come alive with stories. You'll learn why Gandhi hated iodine, why tellurium instigated a bizarre gold rush, and much more.

MIT App Inventor
https://appinventor.mit.edu/
MIT App Inventor is the web-based application development platform Gitanjali used to develop Tethys, her lead detection tool. MIT continues to support and extend this app. The site has information on getting started, tutorials, and teacher resources. If you'd like to create an app, this is an excellent place to start.

MIT Technology Review
https://www.technologyreview.com
Gitanjali says she gets many of her ideas and her technology news from this authoritative source. It's a festival of tech innovations and new ideas.

Rao, Gitanjali. *A Young Innovator's Guide to STEM: 5 Steps to Problem Solving for Students, Educators, and Parents.* New York: Simon & Schuster, 2021.
In her book, Gitanjali relates her own problem-solving method, which combines elements of design thinking with a scientific method. She teaches this process in her online seminars, and this book is interesting and helpful for other young inventors.

CHAPTER 3

The Boy Who Harnessed the Wind
https://www.netflix.com/title/80200047/
This film, adapted from William's autobiography, rightly won critical reviews at its debut at the Sundance Film Festival. In his first time directing, actor Chiwetel Ejiofor focuses on the story of discovery, while honoring the culture and the setting of William's home in Malawi. The film also captures the sense of how William's discovery broke the deadly cycle of famine that had gripped the region. Note: A Netflix membership is required to view the documentary.

"How I Harnessed the Wind"
https://www.ted.com/talks/william_kamkwamba_how_i_harnessed_the_wind?language=en/
William tells his own story in his own words to a TED audience. TED friends played a major role in spreading his story and helping William expand the reach of his ideas and projects.

Make:
https://makezine.com
William says that *Make:* is his favorite magazine. *Make:* is a fantastic resource for people who like to make things: tinkerers, designers, hackers, and DIYers. Each issue has how-to projects, information on tools, and interviews with makers. For example, the November 4, 2020, issue has information on how to build a backyard wind turbine in about eight to sixteen hours for $80 to $150. This project is rated as moderately difficult and uses a bicycle wheel with a dynamo hub, along with good old PVC pipe.

Moving Windmills
https://movingwindmills.org/
Learn more about William's work bringing energy, education, and innovation to Africa and his home, Malawi, on this website. As a part of the Moving Windmills project, William has plans for an innovation center to encourage young African inventors and entrepreneurs.

CHAPTER 4

Dahl, Øyvind Nydal. *Electronics for Kids: Play with Simple Circuits and Experiment with Electricity!* San Francisco: No Starch, 2016.
A great introduction to electronics, the book presents several fun projects that increase in complexity as you read on.

Denworth, Lydia. *I Can Hear You Whisper: An Intimate Journey through the Science of Sound and Language*. New York: Dutton, 2014.
A science writer who discovers her son has hearing impairment writes both about his story and the science of hearing. She interviewed experts on language development, inventors of groundbreaking hearing technology, deaf leaders, and neuroscientists to learn how to help her son.

The Field Guide to Human-Centered Design
https://www.designkit.org/resources/1/
This is a great workbook for people interested in solving tough problems like the grand global challenges. Written by the IDEO field staff, it describes fifty-seven methods to use with social projects along with case studies and worksheets.

Kelley, Tom, and David Kelley. *Creative Confidence: Unleashing the Creative Potential within Us All*. New York: Crown, 2013.
The Kelley brothers, founders of design firm IDEO, show how to tap your own creative talents. They trace the development of design thinking and share anecdotes of design thinking in action. Anyone can draw inspiration here to tackle problems big and small.

Padden, Carol A., and Tom L. Humphries. *Deaf in America: Voices from a Culture*. Cambridge, MA: Harvard University Press, 1990.
Deaf authors Padden and Humphries, using the capitalized "Deaf," discuss the life, complex culture, and the natural, evolving language, American Sign Language, of the Deaf community in the United States.

Platt, Charles. *Electronics: Learning through Discovery, 2nd ed.* Sebastopol, CA: Make Community, 2015.
Platt's book is fast-paced and thorough. It's a good introduction to electronics that progresses from simpler to more complicated concepts and projects.

Stanford d.school
https://dschool.stanford.edu
Stanford's famous d.school site has a wealth of information on design thinking, designing for social benefit, and sustainable design. This is also a good jumping-off point for going deeper to learn about design techniques and design projects.

CHAPTER 5

Howard Hughes Medical Institute (HHMI)
https://www.biointeractive.org/
The HHMI site contains a treasure trove of excellent resources for students and teachers interested in learning more about the biological sciences and current work in biology. This site has both science stories and science activities, such as how to build a paper model of a CRISPR-Cas9 gene editor.

Johnson, Rebecca. *Genetics, 2nd Edition.* Minneapolis: Twenty-First Century Books, 2013.
This book is a good introduction to the history of genetics, beginning with Gregor Mendel's discovery of genes up through biotechnology. The author shows how DNA replicates itself, how genes direct cell construction, and how molecular biologists map the genome.

MIT D-Lab
http://d-lab.mit.edu/
This site describes the amazing work of Amy Smith and her students in developing countries. They develop innovations in areas such as biomass fuels, cleaner cookstoves, and evaporative cooling for food preservation. D-Lab students and staff currently work on field projects and research programs and collaborate in over twenty countries around the world.

Mooney, Carla. *Genetics: Breaking the Code of Your DNA*. White River Junction, VT: Nomad, 2014.
This book, recommended by the National Science Teaching Association, uses graphic novel illustrations and hands-on projects to introduce DNA and genetics. Learn to build a 3D DNA model, extract DNA, use a Punnett square to predict an offspring's probability of inheriting a trait, and more.

National Center for Appropriate Technology (NCAT)
https://www.ncat.org
This site showcases technologies for small farmers and rural homeowners who want lower-cost green techniques. The NCAT provides information, training, and energy audits, and collaborates in the construction of alternative energy sources, such as solar panel arrays.

Science: High School Biology
https://www.khanacademy.org/science/high-school-biology/
This section, by Kahn Academy, explains the basic biology behind DNA, pathogens, and water pollution. The academy is a well-known place for self-paced learning and independent study.

The Water Project
https://thewaterproject.org/
Deepika consulted with and mentioned this nonprofit among others. The Water Project develops projects to provide clean, reliable sources of water to communities in sub-Saharan Africa. They provide training, expertise, and financial support for local development.

CHAPTER 6

Arduino

https://www.arduino.cc/

Originally developed for artists who wanted to make their art interactive, Arduino is an excellent way to get started in electronics for everyone. This site has basic information, tutorials, resources, software downloads, and hardware. Maybe best of all are the kits for purchase that include an Arduino learning guide that takes the beginner step-by-step from very simple projects to more complex ones. These kits include all the parts needed for the projects. For example, Gitanjali made her multimeter from Arduino parts, and the Carl Hayden Community High School team used Arduino to control Stinky the robot.

Ceceri, Kathy. *Making Simple Robots: Explore Cutting-Edge Robotics with Everyday Stuff*. San Francisco: Make Community, 2015.

This book, published by Make: Community, is a good starting point for constructing robots beyond buying a kit. This moves you into the phase where you begin creating and adding your own ideas and flourishes.

Elegoo

https://www.elegoo.com/

There are many good robot kits for people ready to get hands-on experience, and Elegoo offers ones that are easy to build and that work with Arduino. The Smart Robot Car Kit V 3.0 is great for getting started.

Field, Simon Quellen. *Electronics for Artists: Adding Light, Motion, and Sound to Your Artwork*. Chicago: Chicago Review, 2015.

Electronics for Artists, which assumes no prior knowledge of math, science, or engineering, shows how to create projects that light up, move, and respond to the environment. This book is written clearly with many detailed how-to photos and graphics, but it is not a "for dummies" book. Field makes sure to explain *why* things work, as well as *how* they work and how to build them. His two books on Gonzo Gizmos are also great for inventors and tinkerers.

FIRST

https://www.firstinspires.org/

Dean Kamen founded this organization to get teens involved in advanced STEM work while having fun. This organization provides robotics leagues for students from prekindergarten (yes, that's right!) through high school. The site also has at-home activities.

Platt, Charles. *Electronics: A Hands-On Primer for the New Electronics Enthusiast*. 2nd ed. San Francisco: Make Community, 2015.

This is another classic starting point for people new to electronics. It is organized around experiments you can do and things you can make at home. This book is also clearly written and contains many helpful diagrams and photos.

CONCLUSION

Anderson, Maxine. *Amazing Leonardo da Vinci Inventions You Can Build Yourself.* Norwich, VT: Nomad, 2006.
A hands-on book of projects, it includes stories of da Vinci's life and anecdotes about his inventions, along with images from his famous notebooks.

Exploratorium
https://www.exploratorium.edu/
Be sure to try the online features, including Science Snacks and Tinkering at Home. Both of these modules show how to build projects at home that teach important principles of science. If you're ever in San Francisco, visit the Exploratorium science museum for a feast of science. Plan for several hours. It's that good.

"Explore the Top Ten Science Museums in the US"
https://www.nationalgeographic.com/travel/intelligent-travel/2015/10/22/10-great-science-museums-in-the-u-s-a/
Don't live near Boston or San Francisco? The good news is that there are more good hands-on science museums than ever before. You can find museums in most major cities in the United States—do a quick search online to see which ones are near you. *National Geographic* has compiled a short list of some of the best ones in the country.

"14 Grand Challenges for Engineering in the 21st Century"
http://www.engineeringchallenges.org/challenges.aspx
The National Academy of Engineers site lists global challenges and goals.

Grandin, Temple. *Calling All Minds: How to Think and Create like an Inventor.* New York: Penguin, 2018.
Temple Grandin, noted inventor, professor, and person with autism, explores the science underlying various everyday inventions and the ways inventors created and improved upon their ideas. Why does a kite fly? Why does a boat float? Grandin provides clear, simple do-it-yourself projects to show how tinkering can build a deeper learning of these concepts.

Kaufman, Scott Barry, and Carolyn Gregoire. *Wired to Create: Unraveling the Mysteries of the Creative Mind.* New York: Penguin, 2015.
Kaufman and Gregoire explain the newest science of creativity through both research and examples of great creators. They show how to unleash your own creativity in this engaging and accessible book.

Macaulay, David. *The Way Things Work Now.* Boston: Houghton Mifflin Harcourt, 2016.
This updated classic brilliantly reveals the inner workings of devices from windmills (William would appreciate this) to Wi-Fi. Macaulay expanded this concept into a series of books entitled *How It Works* full of illustrated deconstructions of structures from castles to skyscrapers to underground railways. These books provide hours of fascination.

Maker Faire

https://makerfaire.com/

Created by Dale Dougherty, founding publisher of *Make:* magazine, Maker Faire describes itself this way: "Part science fair, part county fair, and part something entirely new, Maker Faire is an all-ages gathering of tech enthusiasts, crafters, educators, tinkerers, hobbyists, engineers, science clubs, authors, artists, students, and commercial exhibitors. All of these 'makers' come to Maker Faire to show what they have made and to share what they have learned." Maker Faire is a global event with local fairs all over the world. If one is near you, it is worth a visit. Each faire is an amazing gathering of STEM people and a mind-expanding exhibition fusing creativity, technical chops, and mad science. It's a festival of technology, tools, and toys for STEM lovers.

Museum of Science, Boston

https://www.mos.org/

The Museum of Science in Boston, like many science museums, offers learning activities online. They also have daily live, interactive demonstrations, such as live animal talks, science of the supercold, and lighting shows, and they produce the Engineering is Elementary curriculum.

"The 17 Goals"

https://sdgs.un.org/goals/

This United Nations site, like the "14 Grand Challenges for Engineering in the 21st Century" site above, lists global challenges and goals for improving the world. Some schools and organizations use these lists to develop curriculum and social action programs.

INDEX

ACKNOWLEDGMENTS

A book is a collective effort and I want to thank the many people who made this all possible. At the start, I was simply following the advice of Toni Morrison, writing the book I wanted to read—and to share and discuss with my science students. Many people, family, friends, and colleagues encouraged me along the way.

I am sincerely grateful to:

- my science students at the Nueva School who inspired me to write this book in the first place, and to curious students everywhere who invent and work to make this world a better place,
- all the teen inventors chronicled in these stories, with a special thanks to Austin Veseliza, Gitanjali Rao, and Oscar Vasquez for their first-person perspectives,
- my extraordinary agent and former owner of the legendary Cody's Books in Berkeley, Andy Ross, who believed in this idea and tirelessly championed it,
- Hallie Warshaw, founder of Zest Books, who saw the possibilities of this book early on,
- all the great people at Lerner Publishing Group, especially Shaina Olmanson, my editorial director, the guiding light for this book; Quinton Singer, my editor, for his creative ideas and his invaluable suggestions; and Kavel Rafferty, our talented illustrator,
- Professor Barbara Kerr, who taught me so much about creative people and the creative process,
- Simon Quellen Field for his astute reading of early drafts and invaluable technical advice, as well as his inspirational gonzo gizmos,
- Sarah Buhre, an extraordinary teacher, for consulting on Deaf culture and hearing impairment,
- Daphne Gray-Grant for unfailing moral support and great writing advice,
- my daughter, Laurel, for ideas, suggestions, and advice, especially on biotechnology issues,
- and my wife, Heather, for her patience, support, and understanding over the life of this project, as well as for reading endless drafts with a keen eye for logical narrative and a poet's ear for language.

Thank you all for your invaluable contributions in this rewarding journey.

PHOTO ACKNOWLEDGMENTS

Image credits: All Canada Photos/Alamy Stock Photo, p. 12; enot-poloskun/Getty Images, p. 28; Bloomberg/Getty Images, p. 36; U.S. EPA Region 5, p. 43; Kathryn Scott/The Denver Post/Getty Images, p. 48; Jeffrey Coolidge/Getty Images, p. 53; Alan Bramley/Alamy Stock Photo, p. 62; Church of emacs/Wikimedia Commons (CC By 2.0), p. 78; Lucas Oleniuk/Toronto Star/Getty Images, p. 82; Richard Watkins/Alamy Stock Photo, p. 97; Rodolfo Parulan Jr./Getty Images, p. 119; grayjay/Shutterstock.com, p. 143; MATE Inspiration for Innovation/MATE ROV Competition/Flickr, p. 157; William Taufic/The Image Bank/Getty Images, p. 172.

ABOUT THE AUTHOR

Fred Estes taught science for nearly two decades in a school near his home in San Francisco. Before that, he taught high school English, worked as a financial analyst, joined an AI startup, developed corporate training programs, and earned a doctorate in educational psychology. Currently, he teaches graduate students and teachers about design thinking, innovation, creative teaching methods, and hands-on science curriculum. He enjoys reading science and science fiction, writing and blogging about science and, best of all, working with small groups of students on science projects. A major contribution of science, he says, was the discovery that dark chocolate contains healthy antioxidants.